George Michael James Giles

**A report of an investigation into the causes of the diseases known
in Assam as kála-azár and beri-beri**

George Michael James Giles

A report of an investigation into the causes of the diseases known in Assam as kála-azár and beri-beri

ISBN/EAN: 9783337220495

Printed in Europe, USA, Canada, Australia, Japan

Cover: Foto ©berggeist007 / pixelio.de

More available books at **www.hansebooks.com**

A REPORT

AN INVESTIGATION INTO THE CAUSES OF THE DISEASES KNOWN IN ASSAM AS

KÁLA-AZÁR AND BERI-BERI

BY

GEO. M. GILES, M.B., F.R.C.S., SAN. SCI. CERT. UNIV. LONDON,
SURGEON, I.M.S., ON SPECIAL DUTY, ASSAM.

SHILLONG:
PRINTED AT THE ASSAM SECRETARIAT PRESS.

1890.

EXPLANATION OF PLATES

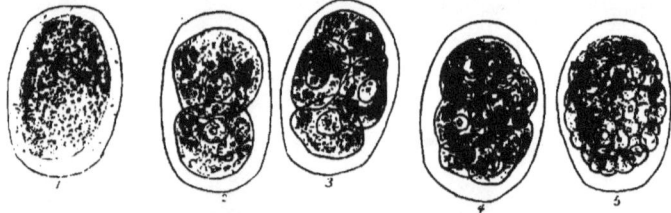

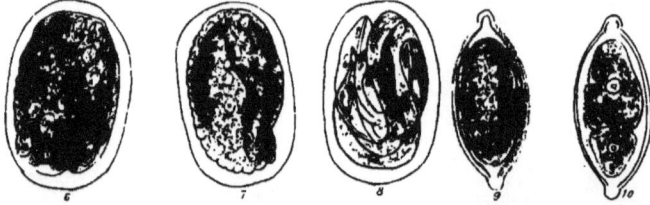

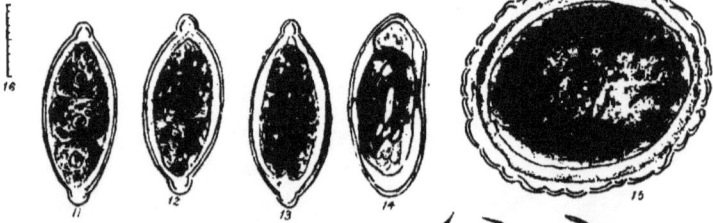

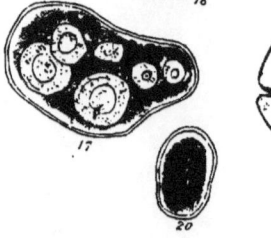

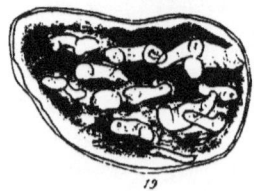

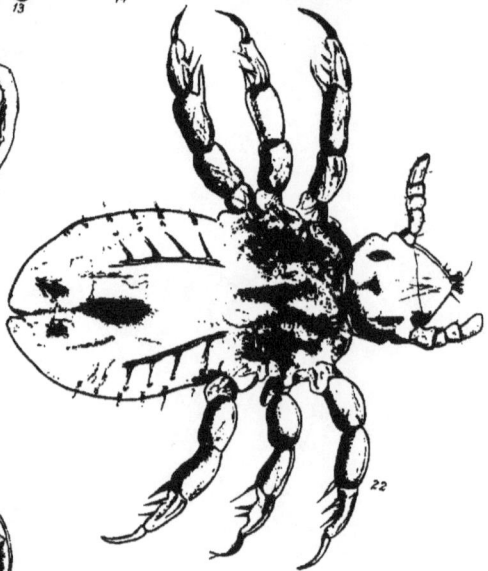

PLATE I.—Ova and other Bodies found in Dejecta.

Fig. 1. Unsegmented ovum of *Dochmius duodenalis* expressed from a dead worm.

  ,, 2.
  ,, 3. } Segmented ova *Dochmius duodenalis* from dejecta of
  ,, 4.   *kála azár* patients.

  ,, 5.
  ,, 6. Subsequent stages of *Dochmius* ova met with in cultiva-
  ,, 7.   tions.
  ,, 8.

  ,, 9. Ovum *Trichocephalus dispar* as met with in dejecta.

  ,, 10.
  ,, 11. Various stages of segmentation of *Trichocephalus* ova
  ,, 12. met with in cultivations.
  ,, 13.

  ,, 14. Ovum of *Oxyuris vermicularis*, as met with in dejecta.

  ,, 15. Ovum of *Ascaris lumbricoides*, as met with in dejecta.

  ,, 16. Line measuring $\frac{1}{1000}''$ drawn to the same magnification as the preceding figures.

  ,, 17.
  ,, 18. Coccidia containing psorosperms from the dejecta of
  ,, 19. *kalá azár* patients.

  ,, 20. A small specimen of the same simulating a *Dochmius* ovum.

  ,, 21. Line of 5 $\frac{1}{1000}''$ drawn to same scale as figures 17-20 inclusive.

  ,, 22. Immature *Pediculus sp.* often found in dejecta.

  ,, 23. Line $\frac{1}{100}''$ long, drawn to same scale as fig. 22.

PLATE II.—The Male Rhabditis of "Dochmius duodenalis" in various Stages.

Fig. 1. Recently hatched embryo.

„ 2. Rhaditis five days old (in the pre-exual stage).

„ 3. Aboral extremity of a male rhabditis in earliest state of sexual differentiation.

„ 4. Hinder half of a male rhabditis further advanced.

    a. Rudiment of testis.
    b. Copulatory spiculæ.
    c. Intestine.
    d. Copulatory bursa.

„ 5. Mature male rhabditis, viewed laterally.

    o. Mouth.
    a. Anterior bulb.
    b. Posterior ditto.
    g. Central ganglion.
    h. Hepatic cells.
    i. Intestine.
    l. Lemniscus.
    t. Testis.
    c. Copulatory bursa.
    s. Copulatory spiculæ.

„ 6. Line representing $\frac{1}{1000}''$ on same scale as preceding figures.

„ 7. Aboral extremity of mature male rhabditis viewed vertically.

„ 8. Line representing $\frac{1}{1000}''$ on same scale as figure 7.

„ 9. Mouth parts of mature rhabditis.

    g. Ganglion cells.
    p. Papillæ.

„ 10. Line representing $\frac{1}{1000}''$ on same scale as figure 9.

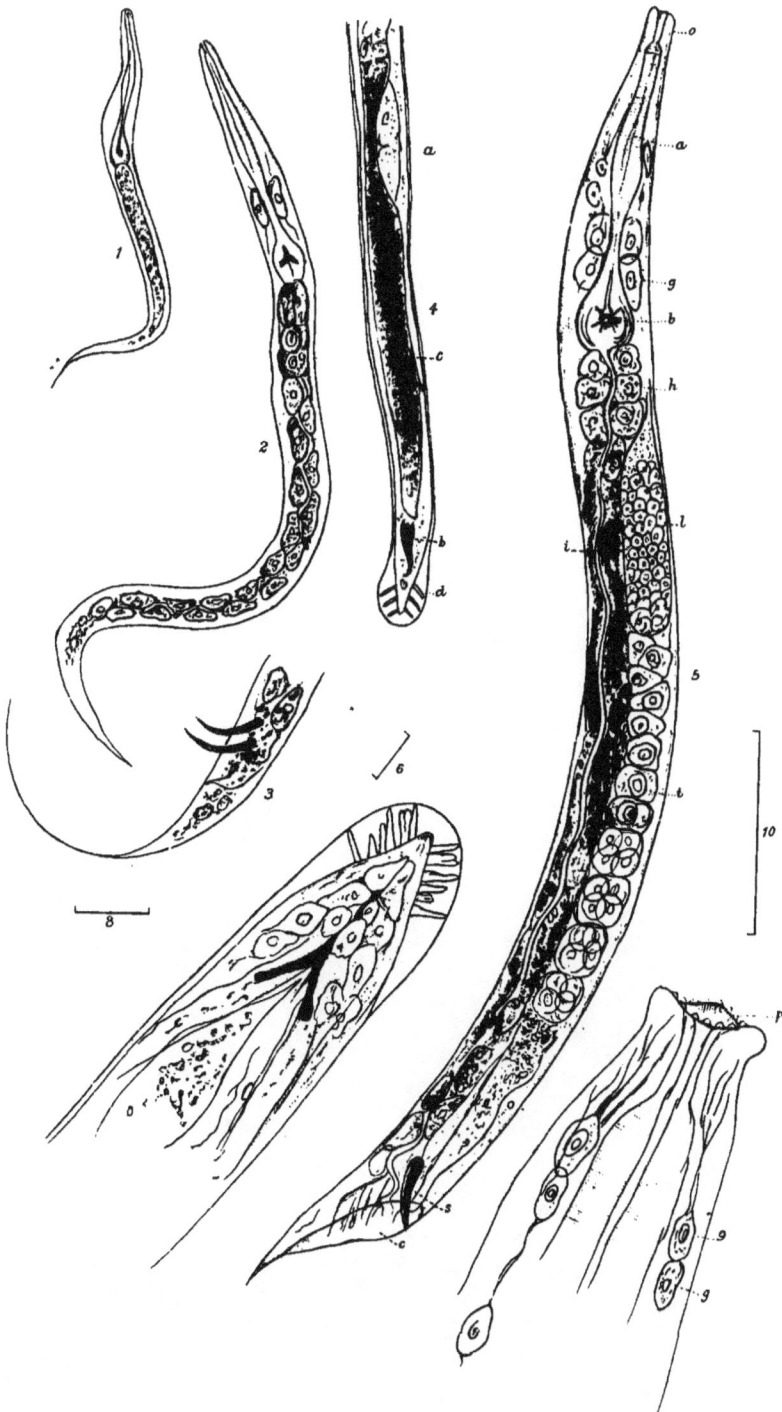

Plate II.

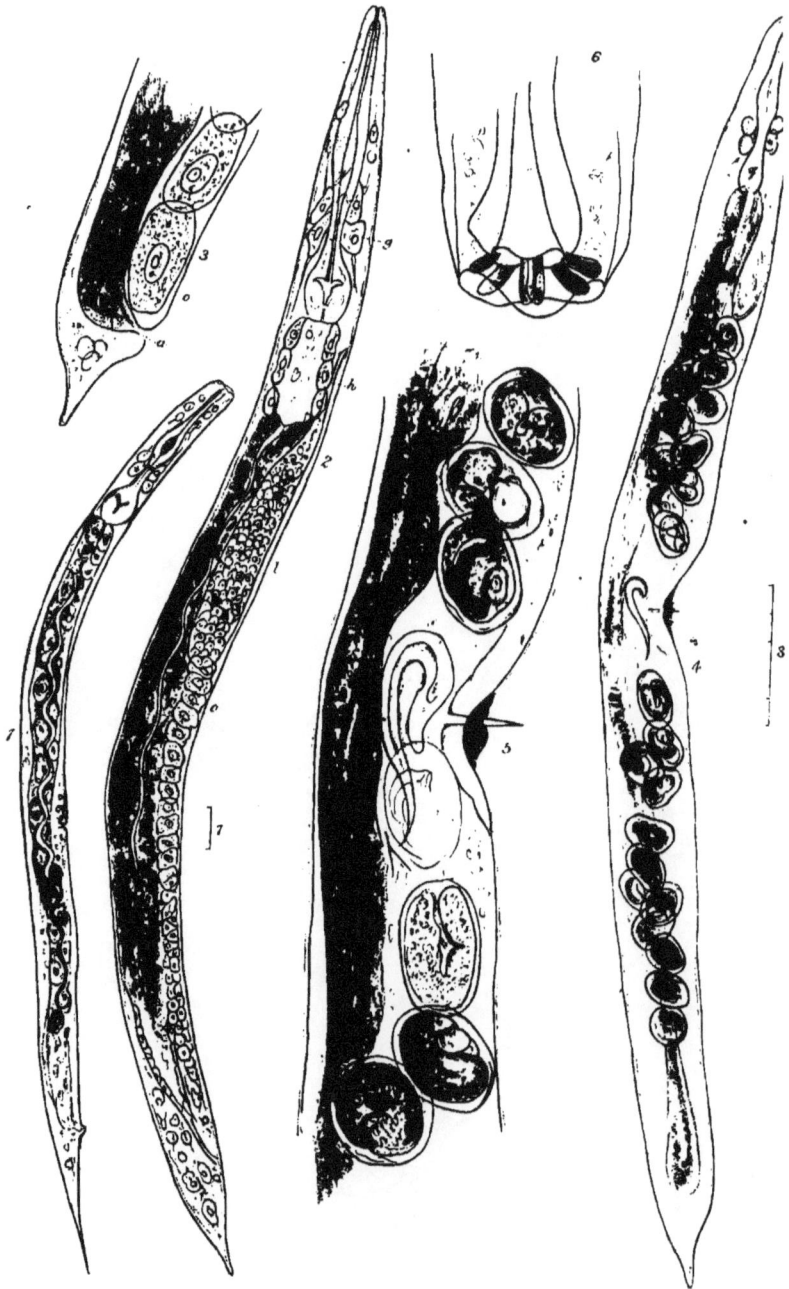

PLATE III.—Stages of Female Rhabditis.

Fig. 1. Female rhabditis in earliest stage of sexual differentiation.

„ 2. Further advanced specimen, the ovary forming a prominent organ, but still histologically indeterminate.

„ ·3. Aboral extremity of advanced female rhabditis showing the blunt extremity of the oviduct containing germs.

„ 4. Advanced female rhabditis almost at the end of the process of depositing her eggs. One embryo has hatched out within her body.

„ 5. Middle portion of the same more highly magnified.

„ 6. Mouth of immature parasitic form from intestine of monkey.

„ 7. Line representing $\frac{1}{1000}''$ on same scale as figures 2, 3, 5.

„ 8. Line representing $5\frac{1}{1000}''$ on same scale as figures 1 and 4.

PLATE IV.—Changes in the Gastro-intestinal Mucosa.

Fig. 1. Portion of a vertical section of upper part of Ilium from
a case of anchylostomiasis, showing the space between
two villi filled up with blood clot, in which is imbedded
an immature *Dochmius* (the worm is cut twice by the
section, so that two sections are seen).   At the highest
part of the recess filled with clot, may be seen a deep
erosion of the mucous membrane. × about 40 diams.

„   3. Vert. section mucous membrane of stomach, thickened,
showing the mouths of the tubular glands blocked by a
fibrinous cicatricial layer. × about 200 diams.

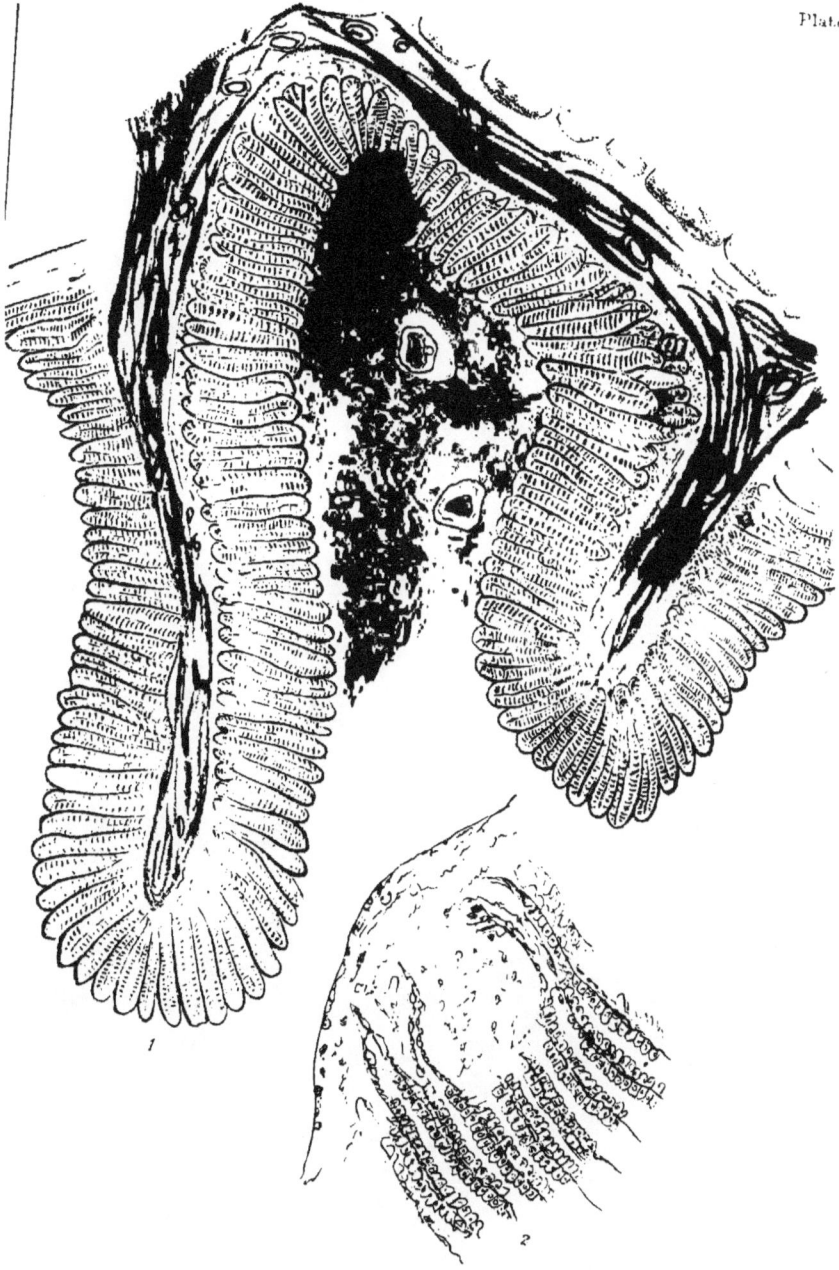

Plate IV

1

2

ABSTRACT OF THE REPORT.

SECTION I.—INTRODUCTORY REMARKS, page 1; the word *beri-beri* has been applied to several distinct diseases, *ib.;* its want of vernacular signification in any language, *ib.;* the diseases described originally by Malcolmson in Madras, and afterwards, by Pekelharing in Achin, are alike epidemic peripheral, neuritis, and have no connection with the *beri-beri* of Ceylon and Assam, which are alike anchylostomiasis, pages 3-7; reference to Leichtenstern's discoveries in Dr. Kynsey's pamphlet on *beri-beri* of Ceylon pages 7-8.

SECTION II.—THE GENERAL COURSE AND RESULT OF THE INVESTIGATION, page 9; preparatory measures, *ib.;* general appearance of *kála-azár* cases, page 10; temperature in *kála-azár,* page 11; its most prominent symptom is anæmia, *ib.;* points of distinction between the anæmia of anchylostomiasis and that of malarial cachexia, page 12; discovery of anchylotomes in a case of *kála-azár,* page 13; hostility of the natives to European medical treatment, pages 13-14; inefficiency of native methods of cooking prevents recovery, pages 14-15; the disease as seen in infected villages, page 15; evidence of the prevalence of anchylostomiasis in *kála-azár* villages, page 16; tour in the tea districts, pages 16-17; negative results of bacteriological investigations, page 18.

SECTION III.—THE PREVALENCE OF KÁLA-AZÁR AS JUDGED BY AVAILABLE STATISTICS, page 19; data are scanty, *ib.;* effect on general vital statistics, pages 19-20; effect on revenue returns, page 21; table of mortality in *kála-azár* villages, pages 22-25; table of villages with a mortality exceeding 200 per thousand, pages 25-26; table of villages with mortality from 150 to 200 per thousand, page 26. The distribution of *kála-azár* is that of a communicable disease, page 27.

SECTION IV.—FACTS RELATING TO THE SPREAD OF KÁLA-AZÁR, page 28. First appearance in 1869, medical relief operations 1884, *ib.;* appearance in Kámrúp, *ib.;* rate of progress of disease, page 29; deductions from the progress of the disease: incompatibility of the facts with the theory of a malarial origin, *ib.;* instances of Paru and Maligaon, page 30; table of analyses of water from localities affected with *kála-azár* and *beri-beri,* pages 32-37; *kála-azár* not necessarily a terai disease, 38; Dr. Russell on the Chaygaon thána, *ib.;* Dr. P. M. Gupta on communicability of *kála-azár,* page 39; further arguments as to the non-malarial origin of *kála-azár,* pages 40-41; has never yet affected Europeans, page 41.

SECTION V.—THE SYMPTOMS OF KÁLA-AZÁR AS ILLUSTRATED BY CASES, page 42; cases taken from Dr. Dobson's reports, pages 43-50; Dr. Dobson on the prevalence of enlarged spleen in apparently healthy subjects, page 50; cases taken from Mr. Macnaught's reports, pages 50-55; cases taken from Mr. Nandi's reports, pages 56-58; cases noted in Charitable Dispensary, Gauháti, pages 59-62; Deputy Surgeon General Clarke on the prominence of anæmia in the symptomology of *kála-azár,* page 62; Dr. Borah on symptoms of *kála-azár, ib.;* Dr. P. M. Gupta on the symptoms of *kála-azár,* page 63; comments on the cases: the dependence of the increased mortality on anchylostomiasis: the inadequacy of malaria to account for the cases, pages 63-66.

SECTION VI.—THE LIFE HISTORY OF THE PARASITE, page 67; early changes in deposits, *ib.;* method of instituting cultivations, *ib.;* period at which embryos hatch out, page 67; measurements of ova, *ib.;* points of distinction between *Dochmius,* and other ova found in dejecta, page 70; changes in ova previous to hatching out, *ib.;* the newly-born embryo, page 71; encapsuling and calcification of embryos, as recorded by Lutz, a fallacy, page 72; method of examination of cultivations, pages 72-73; the process of ecdysis, page 73; changes in embryos while still of indeterminate sex, page 74; meaning of term rhabditis, page 75; the development of the male rhabditis, pages 75-78; the development of the female rhabditis, pages 79-80; reproduction of the rhabditis, pages 80-81; experiments with monkeys, pages 81-85; the question of encystment of the parasite, pages 85-87; duration of life of the parasite, pages 87-88.

SECTION VII.—EFFECT OF VARIOUS CONDITIONS ON THE LIFE OF THE FREE STAGE, page 89; effect of food-supply, *ib.*; if limited, only a single generation is produced, *ib.;* if fresh supplies are furnished, a large number of generations may succeed each other, in the free state, page 90; experiment illustrating these points, pages 90-91; infectiveness of soil may continue indefinitely, page 92; effect of supply of oxygen, *ib.*; the rhabdites can only live on the surface, *ib.*; live but do not flourish in water, *ib.;* deprivation of oxygen kills, page 93; experiments illustrative of this, *ib.;* influence of moisture, page 94; most favourable degree of, *ib.;* can survive moderate dessication of some duration, but not chemical drying, *ib.*; the dessicating influence of sun in dry weather is capable of killing them, *ib.*; the bearing of these facts on disinfection and on the distribution of the disease, influence of heat, page 95; temperatures exceeding 140°F. kill the rhabdites and ova : influence of chemical agents, page 96; influence of light, *ib.*

SECTION VIII.—THE METHOD OF INFECTION BY THE PARASITE, page 97; auto-infection by ova impossible, *ib.;* infection probably only by rhabditis' progeny, *ib.;* drinking water an improbable vehicle of infection, pages 97-98; infected earth introduced into food is probably the usual vehicle, pages 99-102; frequent re-infections are probably the rule, page 102; danger of infection through the agency of milk, page 103.

SECTION IX.—REMARKS ON THE PATHOLOGY, DIAGNOSIS, AND TREATMENT, &C., OF ANCHYLOSTOMIASIS, page 104; importance of early diagnosis and treatment, *ib.*; lesions in the stomach, page 105; the importance of dyspepsia in the production of the fatal train of symptoms, page 106; Cobbold on predisposition in helminthiasis, page 107; Kynsey on the symptoms of anchylostomiasis, *ib.*; the diagnosis of anchylostomiasis by microscopic methods, pages 108-109; Coccidia in dejecta of *kála-azár* patients, pages 109-111; method of searching for expelled parasites, pages 111-112; the importance of systematic medical inspections of coolies, pages 112-113; earth-eating, pages 113-115; remarks on treatment, pages 115-120; complications of anchylostomiasis, other parasites, page 121; *Trichocephalus dispar, ib.; Ascaris lumbricoides,* page 123; *Oxyuris vermicularis,* page 124; *Amphistoma hominis, ib.; Distoma crassum,* page 125; cestode parasites, *ib.;* pedeculi, page 126; intestinal coccidia, page 127; malarial complication, *ib.*, mode of death in anchylostomiasis, *ib.*

REPORT ON KALA AZAR AND BERI-BERI.

I.—INTRODUCTORY REMARKS.

FOR reasons that will appear in the sequel, it will be in no way necessary to enter into any separate description of the maladies indicated by the above popular terms, as the main result of the investigation has been to establish their identity.

It is unfortunate that the word "*beri-beri*" should have acclimatized itself as firmly as it appears to have done in Assam. I refer of course to its use among Europeans, because the word is unknown to the other inhabitants of the province, whether Assamese, or imported Indians. The fact is that "*beri-beri*," the "*kakke*" of Japan, "*kalá-azár*," and another word, equally evil-sounding to English ears, in popular use in South America, but which I am unable at present to recall, all may refer to the same disease ; while each has the added disadvantage of being popularly applied to any disease that may bear a rough resemblance to whatever malady may be the commonest cause of epidemic mortality in the particular region affected.

The confusion has been greatly enhanced by the efforts of scientific men, who were conversant with a disease in some one country, to identify the malady they had themselves observed with diseases of other countries, of which they had no personal experience. The word "*beri-beri*," however, seems to have the widest acceptance, at least among Europeans, but it is a curious fact that it has not even the defence of being in popular use in the vernacular of any part of the world.

On this point Dr. John Grant Malcolmson,* who wrote a prize essay on "*beri-beri*," as observed in Madras, more than fifty years ago, remarks—

" As to the appellation " *Beri-beri*." it appears to me perfectly unaccountable how it could ever have crept into such general use as it has, for

---

* A Practical Essay on the History and Treatment of " *beri-beri*," by Assistant-Surgeon John Grant Malcolmson, Madras Medical Establishment. Vepery Mission Press, Madras, 1835.

B

it is perfectly unintelligible to the natives from whom it is said to have originated. Though the word "B'hay-ree" does really mean a sheep in the Hindustani dialect, and by repeating it (making allowances for the orthographical error in persons unacquainted with the language, and consequently of the proper sound of the word), we may form such a name as *beri-beri;* yet, after the most particular enquiries on the subject, I find the natives are totally ignorant of any disease under that title."

A disease, endemic in Ceylon, of somewhat similar character, came to be called by European medical men by the same name, from the supposed identity of the two maladies, though, it need hardly be remarked, the word has no more vernacular signification in Cingalese than in any other dialect. This endemic, we now know from Dr. Kynsey's pamphlet,[*] is established without doubt to be anchylostomiasis, though it is quite possible that cases belonging to another category may be found in that island in addition. We hear much too of the ravages of "*beri-beri*" in the Dutch Indies, but perhaps here with more justice, if we confine the term to the disease as originally described in Madras. It will be desirable to refer more fully to this point further on, but no one can, I think, compare the descriptions given by Malcolmson[†] and Pekelharing,[‡] without being convinced that they are describing the same disease; and that this disease is not anchylostomiasis, but an entirely distinct malady. At the same time I think it more than probable that in both instances a certain percentage of cases of anchylostomiasis are to be found.

I know, from personal observation on coolies freshly imported thence, that anchylostomiasis does exist, at the present day, in the northern part at least of the Madras Presidency, and it is hardly likely that the disease is at all a recent introduction. Indeed, some few of Malcolmson's cases appear best referable to that cause. On the other hand, Professor Pekelharing actually found anchylostomes in a few cases in Achin. In climates so favourable to the development of the parasite as those of the Madras Presidency and Achin, the spread of the parasite will be merely a question of the habits of the natives, and while I doubt if such habits could anywhere

---

[*] "Report on Anæmia, or *Beri-beri* of Ceylon," by W. R. Kynsey. Government Press, Colombo, 1877.

[†] *Op cit.*

[‡] "Recherches sur la nature et la cause du *beri-beri* et sur les moyens de le combattre," par L. A. Pekelharing et C. Winkler. Utrecht, Kremnik et fils, 1888.

afford more favourable opportunities for its propagation than in Assam, we well know that the sanitation of semi-civilized communities is everywhere so backward that, given the presence of the parasite among such communities, it can hardly fail to find a considerable number of hosts.

Again, it seems possible that at any rate some of the cases included under the name of "*kakke*" in Japan may be examples of anchylostomiasis.

In "Virchow's Archiv.," Vol. XXI., page 290, for 1877, Wernick wrote a paper on the Japanese variety of *beri-beri*, which, as I gather from Pekelharing, was devoted to proving *beri-beri* to be a pernicious anæmia. Pekelharing devotes several pages to demolishing this view, pointing out that anæmia is by no means a necessary, or even common, symptom of *beri-beri* (as he had seen it), never apparently suspecting that he and Wernick may have been observing two entirely distinct diseases, and that their differences might be accounted for by the dissimilarity of what they had to observe rather than to errors of observation on either side. Now, pernicious anæmia is, clinically, by far the most salient symptom of anchylostomiasis.

That the disease described by Malcolmson in Madras is identical with that observed by Pekelharing in Achin is a matter which hardly admits of doubt. Nothing can show this better than to place side by side the cases and symptoms recorded by the two authors. First, let us compare their accounts of the symptoms in general. Malcolmson says—

"The disease presents such a variety of symptoms that it will be more instructive to consider them in detail than to attempt any elaborate general description. It will be sufficient to describe the most remarkable characters.

"It usually commences gradually with a feeling of numbness, sense of weight, and slight weakness and stiffness below the middle of the thighs, sometimes preceded by muscular pains. There is slight œdema of the feet and legs, especially along the tibiæ, often found to come on after the other symptoms. The walk is unsteady and tottering, even when the patient is not aware of weakness in the limbs, which are occasionally tremulous; spasms occur in the calves and soles of the feet, sometimes becoming general and occasionally shooting to the chest and larynx, obstructing respiration and speech. The want of power often rapidly increases to almost total palsy, especially of the extensor muscles, and in a few cases, the patient, after slight indisposition, suddenly loses the use of his legs. Rigidity and various painful affections of the nerves accompany the paralytic symptoms, and there is sometimes pain along the spine, commonly at the two last lumbar vertebræ. In some cases the disease goes no further, and a cure is effected; but more frequently, the numbness extends upwards towards the abdomen, there is a general sense of lassitude and aversion to motion, and the hands, arms, and chest (and in a few cases even the neck and lips) are gradually benumbed. There is oppression and weight at præcordia, dyspnœa on slight exertion, diffused and irregular pulsation in the cardiac region, and the face and hands are puffy and œdematous. The patient is often found dead in bed, or sinks after several fainting fits or throbbings at the heart; or the œdema rapidly increases and extends up the trunk, violent dyspnœa and inability to lie down in bed comes on, with anxiety, cold sweats, cold extremities, rapid feeble pulse, urgent thirst, and partial suppression of urine. At the commencement the urine is always scanty, of a deep red colour, without cloud or sediment, and possessing very peculiar properties; in some old cases it becomes copious, turbid, and pale, with a large white deposit, and is passed with pain from an irritable bladder. The stomach is irritable in many bad cases, and pain and tenderness in the epigastrium is sometimes complained of; there is, in a few, pain in the abdomen, or a sense of heat is diffused over it and the chest. Effusion takes place into the chest, and more rarely into the abdomen, and there are now and then some signs of inflammation of the pleura or bronchi. In the early stage the pulse may be full, hard, and frequent or little altered; when the face is puffy and there is weight and oppression at the præcordia, it is quick, often irregular, and usually small, although it is occasionally strong.

"Various dyspeptic symptoms occur; the bowels are often costive, the stools green and variously disordered, and the eyes are often tinged yellow. The skin is rather cold, unless there is pyrexia, which is often present in the evening. The disease is sometimes fatal in a few hours, but is often chronic, and in these the patient is liable to sudden death, to rapid aggravation of the symptoms, or supervention of new and more formidable ones, by which he is soon carried off; and if he survives these, he may live for a long time bedridden, dropsical, and a true paralytic."

Pekelharing gives the generally described earlier symptoms as follows:—

"At first a slight œdema along the crest of the tibia, the face puffy and doughy, diminished motor power in walking or going up stairs, &c., paræsthetic and anæsthetic symptoms affecting the lower extremities; palpitation of the heart; a pulse somewhat accelerated, or perhaps normal during repose, but which rises to 90—100 on the least exertion * * * * * * Beri-beri always comm ces gradually and without fever, but in a later stage there may occur a sud en exacerbation of the malady, accompanied with fever * * * * * * An acute exacerbation may bring about a sudden termination of the case, from the nerves of the heart becoming involved, or become the cause of dropsical complications. Beri-beri does not always commence with cramps, but, during a certain stage, an acute exacerbation affecting a large number of motor nerves may be accompanied with the phenomena of irritability of the muscles. At first paræsthesia is not a prominent symptom; but an acute exacerbation may be accompanied by neuralgic pains, and cause the terrible suffering known as anæsthesia dolorosa."

Both authors, too, remark on the circumstance that residence within an endemic area of some duration is necessary before the disease can definitely manifest itself, and are equally agreed as to its depending primarily on nervous lesions. The comparatively rude methods of clinical and pathological investigation available fifty years ago naturally prevented Malcolmson's recognizing the peripheral origin of these lesions, but he describes clearly enough, from advanced cases, lesions of the spinal cord and larger nerves, which we know now to be of a secondary character, and entirely similar to the advanced lesions described by Pekelharing. Compare, too, the following two cases, extracted, almost at random, one from each author.

The case I extract from Malcolmson is reported as follows:—

" James Hicks, Indo-Briton, April 24th, for some time ill of *beri-beri*; limbs cold, moist, and numb; œdema of back and legs, partial loss of power of lower extremities; tendons in hams sometimes contracted, and legs are extended with difficulty; spasms of calves; urine pale, depositing earthy phosphates. 26th.— Heat of feet nearly restored; numbness much the same. 27th—Feet mostly warm, when they sweat they get cold. May 4th— Less numbness, and cannot bear the heat of sand in which he has been walking at noon for some days; coldness of toes and heels only, cold sweat of forehead. 19th—Legs to-day cold and covered with sweats. Vespere. Skin very hot all day. Pulse 100. Thermometer in axilla and palms of hands 101°; between soles of feet, which, like the rest of the benumbed parts, do not feel hot, 77°, and above the knees a little higher. Temperature of the air 89°. Barometer 30'. 20th.—Fever left with sweating. Temperature of axilla and hands 98°, feet 91°, and says they are not so cold as before. Temperature of the atmostphere 86°, Barometer 29' 58". Hardness of the flesh over the lower part of calves, and a cold sweating covers the part. Internal feeling of heat in right leg. Says the 'veins are now getting a colour from the blood, and that before they were empty.' After this, while the temperature of the air varied from 80° to 98° (Barometer 29' 48"), the feet were of the same temperature as the air, and the hands 97,' but when that of the air fell to 83° and it was loaded with sensible moisture, the feet were at 87° and the hands at 94°. The numbness was not complete, and slight œdema extended apparently into the interstices of the muscles. In the progress of the case, burning sensations in the feet and calves of legs came on, but the surgeon into whose care he passed did not ascertain the temperature. The arterial action did not seem to have been weakened in the cold parts, but the colour of the blood in the veins, as far as could be judged from that of the vessels, was not of the usual dark hue."

The case extracted from Pekelharing is that reported on page 23 of his paper:—

" Mang, aged apparently about 21, born at Batavia, has served for four months as a sailor on board the *Hydrograaf*, off the island of Onrust. There were several cases of *beri-beri* on board. He had never been ill before. According to his own account, which there is no reason to doubt, he became ill on 4th December 1887. In a few days his legs became

swollen ; next, he experienced an unbearable tingling, but what occasioned
him the greatest suffering was acute pain in the calves. He described this
pain as being in the bones. In the meantime movement, and especially
walking, had become difficult. On the 13th December he was admitted
to hospital at Batavia. The swelling of the legs diminished rapidly, the
parœsthetic symptoms lasted a few days longer; the pain disappeared as
long as he remained quiet, but re-appeared if he attempted to walk. He
was not quite paralysed, and never had been, but his walk was tottering.
 " The patient stated that he had never had any fever, nor had he had
any difficulty in obeying the calls of nature ; he ate, drank, and slept well.
 " Condition on 31st December.—The patient is strongly built and muscular, and does not look as if suffering. There is no apparent anæmia or
blueness of the mucous membranes. He enters with interest into what
is going on around him, and describes his condition from the commencement of his illness with considerable vivacity. Pulse 94, rising to 116
after very little exertion, moderately tense ; neither sharp nor dichrotic.
Respiration 26, of abdominal type. The heart's impulse is strong, and can
be made out as far as the 5th intercostal space in the nipple line. The
heart's dullness extends beyond the left border of the sternum, but does
not reach the middle line. The sounds of the heart are sonorous, and
pure, with reduplication of the second sound and intensification of the
pulmonary disastolic sound.
 " No other abnormalities could be discovered in either the thoracic
or abdominal organs. His face is doughy and somewhat puffy, but
does not pit on pressure. Along the crest of the tibia the skin
pits distinctly, but the ankle is not œdematous. The pupils act promptly,
and evenly, both to direct light and for convergence. The higher senses are
unaffected. He sees well, hears the tick of a watch at a long distance, and
distinguishes at once between salt and sugar. Organs of motion.—The
muscles are well developed, hard to the touch. Some of them, for example,
those of the calf, are very tender on pressure, and seem to have become more
resistant ; when they contract, one meets with circumscribed swellings ;
mechanically, they were not irritable to the stroke of the percussion hammer.
They present no ideomuscular contraction.
 " Most movements, whether active or passive, can be performed.
However, when the patient is on his back, he cannot flex the ankle, and
plantar flexion can be easily prevented.
 " If the patient rises, considerable locomotor disturbance is observable.
He lifts his foot with difficulty from the ground, raises it high ; he cannot
stand upon one leg, nor raise himself on his toes. If he try to crouch,
he does so clumsily, and is in danger of falling. When lying on his back
he can bend and extend the trunk, and perform all movements of the upper
extremities, as well as those of the eyes, face, and tongue. When
standing with closed eyes, he totters, but no distinct disturbance
of muscular coordination can be made out. · Lying with the eyes closed,
he can place the toes of one foot on the other knee slowly, but without
blundering."

Allowing for the different ways of describing similar facts
incidental to the advance in the methods of medical investigation that has taken place during the interval of more than
fifty years which has elapsed between the dates at which the
two descriptions were written, I do not think that any one can

doubt, after reading the above extracts, that both authors are describing the same disease.

Now, this epidemic affection of the nervous system presents not the least similarity with anything met with in Assam. Compare these accounts with the symptoms of anchylostomiasis as given in Dr. Kynsey's pamphlet (page 5, paragraph 26), which, being readily accessible to all interested in the diseases of Assam, I will not occupy unnecessary space by extracting. In the one disease the prominent character is paresis, and anæmia is an after and unessential complication ; in the other, the leading symptom is anæmia, and no true paralysis ever occurs.

Seeing then the confusion that has been caused by the use of the word, and the fact that it has nowhere any vernacular signification for the patient affected, but is used only by Europeans, it would, I think, be in every way preferable to entirely drop the use of the word and to adopt some such words as anchylostomiasis or parasitic anæmia for the *beri-beri* of Ceylon and Assam, and specific peripheral neuritis, or endemic palsy for the disease described by Malcolmson and Pekelharing in Madras and Achin respectively.

Before closing these introductory remarks, I would wish to point out that, situated as I am at the time of writing, far from all medical libraries, it is quite impossible for me to do justice to those who have worked at the question before me, or to attempt to give any bibliographical references whatever.

I have not even been able to gain access to Leichtenstern's paper, briefly referred to in Dr. Kynsey's compilation, though it is possible that he may have followed the *rhabdites* already described by many previous observers to the adult condition, in which case much of my work may have been forestalled by him. However, not having seen the paper, it was necessary to work out the whole question anew, and I am necessarily ignorant of how far his results go, or to what extent they coincide with my own. There must, however, be some mistake in either the original paper or its translation. The paragraph to which I refer is to be found near the bottom of page 52 of Dr. Kynsey's pamphlet, and is worded as follows :—

" The work of Leichtenstern announces the discovery of a rhabditic form, the segments of which have an independent life when separated, and are able to propagate the species."

Now, to speak of the "segments" of a nematode worm, which is not a segmented animal in either the biological or the popular sense of the word, is a zoological absurdity, as egregious as would be the description of a vertebrated molusc or an eight-legged insect. The description would apply well enough to a tapeworm, although it would hardly be, strictly speaking, correct to speak of the proglottides of a cestode as "segments," a term which, in biology, is properly restricted to the externally visible somites of arthropods, but, taken as it stands, the statement is utterly unintelligible. If we substitute the word "individuals" for "segments," and strike out the words "when separated" altogether, it becomes not only intelligible, but in accordance with the facts of the case as I have observed them; but it is difficult to understand how even the most careless translator could make such a combination of mistakes, unless, indeed, he mistranslated the word corresponding to "segments" and, on his own authority, inserted the words "when separated" in order to explain it.

On this account I have been much disappointed that, up to date, my correspondent in England has not forwarded me the original paper. It is, of course, possible that the note in question may be a translation of an abstract originally made by an unskilled hand, or may even have passed through several hands before assuming its present form; but, in the absence of the original, it is difficult to say where the very obvious errors took their origin.

Before coming to Assam, I took the opportunity of searching for references to *beri-beri* and anchylostomiasis in a file of the Medical Record, but came across very few. Those that I found referred exclusively to cases of *beri-beri*, which come under the heading of endemic peripheral neuritis. My file did not, however, extend back for more than a few years.

II.—The General Course and Results of the Investigation.

Léaving the Central Provinces on the 1st November 1889, Calcutta was reached on the 4th. Though well provided with microscopes and microscopical accessories, I possessed no bacteriological plant, apparatus of this kind being too bulky to form part of the kit of a medical officer engaged in his usual routine work.

In view of the fact that the cause of *kála-azár* was quite unknown while the researches of Pekelharing and others indicated a bacterial origin for *beri-beri*, it was clearly necessary to be provided with sufficient bacteriological apparatus to start laboratory work of that kind. I accordingly obtained permission from Deputy-Surgeon-General C. P. Costello, F.R.C.S., Sanitary Commissioner of Assam, to halt a sufficient time in Calcutta to procure what might be necessary. Apparatus of this kind is of course not kept in the very limited stocks of the purveyors of scientific appliances in any Indian town, so that it was necessary to get many of the requirements made, and some other articles, through the kindness of the Superintendent, Dr. Wood-Mason, were purchased from the Laboratory of the Indian Museum. I also chose at the Telegraph Maintenance Company's establishment a medical battery, in case I should afterwards require one. The disease observed by Pekelharing in Achin depends essentially upon nerve lesions, and for its investigation electrical apparatus is absolutely necessary; but, from what I had already read, it appeared tolerably certain that the disease described by him was quite distinct from that known by the same name in Ceylon and Assam, so that in all probability it would be unnecessary to spend Government money in the purchase of one. As, however, in no instance have cases been met with exhibiting symptoms pointing in any way to nervous lesions, it has never been necessary to buy the apparatus chosen. While such of the articles were being made as required personal supervision, I employed myself in preparing a stock of tubes of cultivating media, which I was enabled to do much more quickly and easily in Calcutta than could have been done elsewhere, because there, by the courteous permission of Dr. Warden, the Chemical Examiner to Government, the resources of the well-equipped laboratory of the Medical College were entirely at my disposal.

Owing to the unavoidable delay so caused, it was not until late in November that I reached Gauháti, which had been indicated as the best point from whence to commence operations. Here I found a most zealous and obliging coadjutor in the person of Dr. Mullane, the Civil Surgeon, who not only set aside for my use a room in the dispensary, but found accommodation in, his house for myself and laboratory, but for which, owing to the entire absence of other suitable accommodation, it would have been very difficult to proceed with the work at all. There were five or six in-patients in the dispensary suffering from *kála-azár*, as well as a daily casual attendance of out-patients suffering from the disease, and the first step was clearly to make a careful examination of the cases at my disposal.

The greater number certainly presented more or less prominent symptoms of malarial poisoning, but it was equally apparent that, in by far the larger proportion, the malarial symptoms were quite inadequate to account for the gravity of the mischief. On casually enquiring the history of a case, the patient would generally say that it had commenced with " fever," and he had it off and on for months. On a more and more close interrogation, however, so as to distinguish true ague from other maladies, it generally came out that there had been comparatively little true fever, and that what the patient really meant was merely that he had been feeling ill for a long time. There is of course nothing new in this, because everywhere in India, nearly all disease is ascribed by natives to " fever," and it is only by the most patient enquiries as to the exact symptoms actually experienced that one can get any other history for nine diseases out of ten.

Then, again, most of the cases had more or less enlargement of the spleen, many of them exhibiting a regular " ague cake ;" but when we remember how rare it is in *post mortem* examinations to find any Indian with a spleen of the normal weight, it will readily be seen that too much importance should not be attached to the presence of a symptom which is well nigh universal, and is by no means incompatible with fair general health. A much enlarged spleen may of course do damage by its pressure on the other abdominal organs, but otherwise it is important only as a sign of malarial poisoning. It is that poisoning, and not the splenic enlargement caused by it, that is the cause of malarial cachexia, for we know that the spleen can be entirely removed without, to all appearance, affecting the general health. Hence it is a mistake to attach any great importance to an enlarged spleen, unless its presence be accompanied by other symptoms of malaria. It is a matter of common experience to find men doing their full day's work, and quite healthy to casual examination, and yet having a spleen so large that it may extend down to the iliac bone and across the middle line. Such subjects are in constant danger of death from rupture of the organ, caused by comparatively slight injuries, but otherwise do not appear much worse off than their fellows. Indeed, spleens as large as this may be found hidden under the folds of fat covering sturdy Banniahs, whom no one would consider cachectic.

Now, the cases in the dispensary, though in a terrible state of cachexia, gave no other evidence of malarial poisoning than a very variable amount of splenic enlargement. The temperature charts, far from forming the well-known malarial trace, exhibited, as their most marked character, a subnormal temperature, indicative of a profound depression of the vital forces. At first I could not, and would not, believe the thermometric observations to be correct, but repeated and careful observations with several thermometers which were tested by comparison with thoroughly reliable instruments of non-clinical make, showed that this low temperature was an actual fact in all advanced cases.

I have observed the temperature as low as 94°F., and this in cases by no means absolutely dying. A temperature of 95° rising to 96° in the afternoon was often persistent for several days together; such low temperatures are almost unknown except in the case of profound traumatic shock or in *articulo mortis*, and are certainly unknown to persist for any length of time in any other disease than anchylostomiasis. This and the profound anæmia were the most marked clinical characteristics of the disease with which I had to deal.

Putting aside cases of ordinary sickness of all sorts, which were freely brought me as *kála-azár*, this anæmia was the one constant symptom. As has been repeatedly noted by previous observers of *kála-azár*, it is the earliest symptom to appear, and its intensity advances *pari passu* with the disease. Its characteristics easily distinguish it from the anæmia that accompanies malárial cachexia. In the latter, anæmia is a late and secondary symptom, whereas here it appears from the first, and nearly all of the other symptoms are merely its results. Then, again, in the anæmia of malarial cachexia, specially when accompanied by enlarged spleen and consequent venous and portal obstruction, the conjunctiva, though deficient in blood, is nearly always of a dirty-yellow tint, often accompanied by distinct icterus. The absolute dead white, rather blueish than yellow, coloration, met with in the larger proportion of *kála-azár* cases is never met with in uncomplicated malaria. There is something peculiarly pathognomonic about this appearance of the conjunctivæ which, once thoroughly appreciated, is not easily forgotten.

At this time, however, I had no idea as to what might be the cause of the symptoms observed, and, while awaiting an opportunity of making a *post-mortem* examination, proceeded to make a series of cultivation experiments by inoculating a considerable series of tubes of nutrient jelly with blood from several patients. With the exception of a few tubes, which developed accidental colonies of well-known mildews, &c., such as will occur in any such series of observations, these experiments gave entirely negative results. These observations, and the fact that none of the cases showed the least sign of paralysis, or disturbance of sensory power, clearly showed that, whatever it might be, *kála-azár* had no connection with the endemic palsy described by Pekelharing under the name of *beri-beri*. I may here mention that neither in *kála-azár* nor in the *beri beri* of coolies have I met with any truly paralytic symptoms or any trace whatever of nerve lesions. Hebetude and an excessive weakness which may simulate paralysis, are of course the common characteristic of the last stages of these maladies, as they are in other similar states of low vitality, but it would be as erroneous to speak of this state as paralytic as to so characterize the weakness of a man tired out by severe exertion, for it depends, not upon any lesions of the nervous system, whether central or peripheric, but upon want of nourishment for brain and muscle.

After about ten days, the death of one of the in-patients enabled me to make an autopsy. The examination showed that the immediate cause of death was an exacerbation of a state of chronic dysentery from which she had suffered during the whole of the time she had been under observation. The other changes were œdema and ascites, excessive anæmia of

all tissues, and an obviously thin and watery condition of the
blood. In the duodenum and upper part of the jejunum
were a number of anchylostomes. In this case, then, the primary cause of death was plainly enough anchylostomiasis,
and this led me to examine the dejecta of my other patients,
and in every instance enormous numbers of the ova of the
parasite were found. A rough estimation was made of the
numbers passed in a few cases by diluting a known weight
of fæces, and counting the number of ova in a small weighed
portion of the diluted material, with the result of showing
that the number passed daily must often exceed a million.
Other *post-mortem* examinations followed, and proved
incontestibly that, whatever *kála-azár* might be elsewhere, the
disease so called in Gauhâti was undoubtedly anchylostomiasis. Such being the case, my next step was to gain
some preliminary familiarity with the appearances of the free
embryonic forms of the parasite, and, while still observing
such cases as presented themselves at the dispensary,
sufficient sets of preliminary cultivations were instituted to
ensure the ready recognition of the embryoes wherever they
might then be encountered. It may be well here to anticipate
an objection which might well be raised, namely, that embryonic
nematodes of different species resemble each other so closely
that one might readily confound rhabditic anchylostomes
with *anguillulæ* and other nematodes found commonly enough
in the soil. The characters, however, of the bursa of the male
rhabdites are quite sufficient to distinguish our species from
any other nematodes likely to be met with in such situations.
Having made these preparations, I went into camp, proceeding
first to the Chaygaon district, which had been pointed out as
the most favourable base of operations. I visited a large
number of *kála-azár* villages, but was much disappointed
in the opportunities they offered for clinical observation.
I saw, of course, crowds of cases, but the people could not be
persuaded to come to the dispensary for either treatment or
observation. There is a branch dispensary at Chaygaon, and
there must be hundreds of cases of *kála-azár* within a radius
of a couple of miles of the building; but, in spite of this, only
three or four cases attend daily, and the number of new cases
was quite insignificant; and even if one carried a supply
of medicine with one, it was difficult to induce the sick to take
it. They have such a firm belief in the incurability of the
malady, that they are free to confess that even charms and
sacrifices to the gods are quite unavailing.

The remark recorded by Dr. Dobson of one of his patients well illustrates the position they take up on this point.

He was trying to persuade an old woman to submit to treatment by impressing upon her that *kála-azár* was but neglected fever, the view of the malady which has hitherto been universal with the profession. Replied she—" If I have got fever I shall get well sooner or later anyhow, but if I have *kála-azár* nothing can cure me, so what is the good of bothering me with medicine ? " From the point of view I take of the disease, there was a great deal of justice in the old lady's remarks. Nothing that has been hitherto tried has had any effect on *kála-azár*, and even *thymol* is as a rule of no use, because cases seldom if ever come to us sufficiently early to be relieved of the parasites before they have inflicted fatal damage upon the system.

It is little use stopping the drain of blood from the patient's veins, when his digestive powers have become so utterly wrecked as to be quite unable to assimilate food to replace the loss. The inefficiency in advanced cases of measures merely directed to expel the parasites is only too well known to all tea-garden medical officers, who have much experience in the treatment of anchylostomiasis. Even in early cases, they are, I find, agreed that the expulsion of the parasites is but the first step of the treatment, and is efficacious only as a preparatory measure to careful feeding up on the "hotel system." An advanced case rarely recovers, whatever one may do. It is just possible that such cases might be conducted to recovery, could one place them in the wards of a London hospital and feed them hourly with peptonized liquid nourishment, but such measures are quite impracticable amongst Indians, whose caste prejudices forbid their touching any food save such as has been prepared by a caste-fellow, even if one had the appliances for providing anything more suitable. Even in our dispensaries, the cooking is of the rudest, and, amongst themselves, no native has the faintest notion of sick-room cooking. What is food for the healthy must serve for the sick also, and he who cannot digest half-cooked *dal* and ill cleaned, sodden rice must necessarily die.

In the course of this investigation I must have microscopically examined the dejecta of many hundreds of subjects, sick and healthy, and nothing impressed me so much as the inadequacy of native methods of cookery which it brought to light. The rice is commonly so badly cleaned that one finds, on washing away the finer particles, an enormous quantity of whole rice-husks. Now this is a very different matter from the comparatively finely-ground particles of husk to to be found in our own brown bread, yet even this is so irritating that the article has a well-known purgative action. How much more irritating then must be the unbroken husk of rice, which is more siliceous than that of any other grain, and has extremely sharp points. In the case of *dal* it is common to find it passed quite undigested, the cooking having quite failed to soften the grain, which, if thoroughly cooked, is one of the most digestible of foods. In jails, where we put into practice on a large scale these imperfect native plans of cookery, which at the best are only fitted to deal with small quantities, this is even more marked than among the free population, and is, I strongly suspect, one of the most important factors in causing the disproportionate amount of bowel-complaints found in such establishments.

However, although it was quite impossible to make a very minute clinical study of the disease, there was no difficulty whatever of satisfying oneself of its enormous prevalence, for the people were ready enough to bring out their sick and to converse about the disease, when one visited them in their villages, however unwilling they might be to adopt our methods of cure.

By making daily excursions from Chaygaon, a large number of villages were visited, in some instances half depopulated, in others with only a few cases, and while, as will be more fully pointed out further on, cases of sickness of all sorts were confounded with the epidemic cause of additional mortality, by the panic-stricken villagers, it soon became evident that by far the greater proportion were cases of anchylostomiasis, and that it was this alone that was responsible for the enhanced mortality.

One strongly confirmatory fact was that large numbers of the villagers who did not as yet consider themselves as absolutely ill, showed unmistakable symptoms of the disease, and, after a little practice, it became easy to pick out from the group around one, a number of such cases at a glance.

The diagnosis of the disease can only be made a matter of certainty by the discovery of the ova of the parasite in the dejecta, and even this test cannot be considered final until after careful and repeated examinations.

It was, of course, impossible to induce the people to bring specimens of their dejecta except in the very small number of cases one met with in hospital. Owing to this, it was found necessary to test the question of the prevalence of the disease by collecting specimens of the dejecta of the inhabitants at random ; but as they never go more than a few yards from their own doors to perform the offices of nature, there is no difficulty whatever in finding specimens, for, filthy as are Indian villages in general, I never elsewhere have found the people so thoroughly careless in this respect as in Assam.

Close by the houses are always to be found a number of small, shallow pits, from which mud has been taken for plastering the walls. As the Assamese has a constitutional objection to unnecessary exertion, these excavations are never more than 10 or 20 yards from the house, and they are fond of utilizing them as latrines, forming miniature, but horribly offensive, open cesspits close to the dwellings. Of course, there is much fouling of the general surface, as they will often defile the ground absolutely beneath .the eaves of their huts, to save themselves going out in the rain, and are too utterly careless about the matter to use any one place systematically, but these pits are certainly their favourite place for the purpose. By examining thus specimens taken at haphazard, it was often found that, in badly-stricken villages, three specimens out of every four would contain the ova of the parasite. Further, it was found that the severity of the outbreak and the proportion of specimens showing ova was generally proportionate. In these pits, in the soil about the houses (though often showing no signs of recent defilement) in the puddles in the streets, and in the filthy shallow *bhils*, I repeatedly discovered the free or *rhabditis* phase of the parasite, in all stages of growth. Strange to say, however, in only one instance has a trace of it been discovered in any specimen of ostensible drinking water, and even this instance was a doubtful one. The worse a village is affected with *kála-azár* the more easy is it to discover evidences of the prevalence of the parasite.

After some stay in the *kála-azár* districts, I proceeded to Upper Assam, and visited the tea-growing districts, where anchylostomiasis has recently been shown to be so prevalent under the name of *beri-beri*.

Nothing here struck me so strongly as the absolute identity of the clinical pictures presented by these cases of acknowledged anchylostomiasis with those I had just been seeing so much of, under the name of *kála-azár*. Probably, one of the reasons that has prevented the earlier recognition of their identity is owing to the fact that no one medical officer has had any very extensive opportunities of observing both diseases. Goálpára, *e.g.*, has been continuously under the charge of Dr. Dobson, but that officer has never served in the tea districts, while the tea cultivation about Gauháti is too limited to afford much opportunity of studying the diseases of coolies to the Civil Surgeon there. Surgeon-Major Borah, on the other hand, has had comparatively little opportunity of observing *kála-azár*.

Owing to the very extensive drainage operations that have been carried out on the level tea lands in the Lakhimpur district, the malariousness of the localities inhabited by the garden coolies must necessarily have been much diminished, and they are much less obnoxious, in this respect, than any part of either Kámrúp or Goálpára; and for this reason, alike among the victims of anchylostomiasis, and among the apparently healthy, enlargement of the spleen is much less commonly met with. Nevertheless, it is easy to find numbers of cases, which exhibit as large spleens combined with anchylostomiasis, as in the worst cases of *kála-azár*, and no one, in these cases, appeared to consider the malarial complication as of other than secondary importance.

Probably, we shall never hear of *kála-azár* in the tea districts about Lakhimpur, because the cases will be at once identified, and it will be said that *beri-beri* has spread to the indigenous population, as, indeed, it already has, in the instance of some villages near Dibrugarh. For the same reason, when anchylostomiasis appears in the tea-gardens of the Kámrúp district, it will be called, not *beri-beri* but *kála-azár*,* and, as in that district, the deep drainage cuttings found in the tea-gardens about Dibrugarh are not required, these will no doubt be found to possess as large a proportion of enlarged spleens as is found among the ordinary population.

What I found most noticeable in my tour in the tea districts was that, while a great deal of money had been spent upon improving water-supply, the provision of suitable lines, and other sanitary projects, the matter of conservancy was

---

* It is a curious circumstance that, since the above lines were written, the forecast therein made has proved to be a true one in the instance of the Chunsáli tea-garden near Gauháti, where, though the management returned the mortality as due to " anæmia ;" the Civil Surgeon describes the outbreak as " *kála-azár*," exactly like that met with in the district, *vide* Section XI.

everywhere entirely neglected. It further appeared that, while these improvements have undoubtedly resulted in a considerable improvement in general health, as far as anchylostomiasis is concerned, they seem to have been generally ineffectual. These tours, including a week's visit to the immigration depôts at Dhubri, which resulted in the discovery that the disease is already present in $2\frac{1}{2}$ per cent. of the newly-arrived immigrants, occupied my time fully until the end of March.

Having thus satisfied myself that, in both diseases, the mortality was alike due to anchylostomiasis, my duty clearly resolved itself into making a careful investigation of the life-history and method of infection of the parasite. A great deal was already known, but many points were still doubtful, more especially it was important to ascertain the best way of destroying the infective embryos.

This task has occupied the greater part of my time during the remainder of the period, but another, and very tedious, business was the systematic examination of the pathological material that had been collected. With the view of supplementing the cultivation experiments made with the blood of *kála-azár* patients, and which gave purely negative results, I made and examined, under a high power, immersion objective, many hundreds of sections of the organs of the subjects of *post-mortem* examination. These sections were stained in every possible way likely to reveal the presence of *bacteria*, but without bringing to light anything specific, though, in cases that had died of lung complications, I found the now often described *cocci*, and made some other observations of a similar character. Had any specific *bacterium* been present, it could scarcely have escaped notice. The greater part of each day, for over two months, was occupied in this work, so that the examination was in no sense perfunctory.

This result forms a strong piece of negative evidence in favour of the view that anchylostomiasis is the one and only cause of the enhanced mortality, for nearly all epidemic diseases have one by one been shown to be due to the action of parasitic organisms, vegetable and animal, and, where the cause cannot be shown to be due to the action of vegetable parasites (*bacteria*), the probability must be in the direction of an animal parasite, so that, wherever we meet with an epidemic of hitherto unknown character, we may fairly expect to be able to refer it to one or the other category.

The examination of old records, and the preparation of this report have also necessarily occupied a considerable time.

III.—The Prevalence of *Kála-azár* as Judged by Available Statistics.

That the mortality from *kalá-azár* has been very serious, is a matter about which there can be no doubt, but unfortunately it is quite impossible to gain any exact idea of the absolute number of deaths it has caused. It is certain that it has, in many cases, nearly depopulated whole villages, and caused such panic among the survivors that they have fled and left the place entirely abandoned. Several places were pointed out to me during my tour in the Chhaygaon district, where, in some instances, the ruined habitations of families, which had been practically exterminated by it, exceeded the number still inhabited. In such instances, enquiries among the surviving inhabitants would generally elicit the tale that most of the inhabitants of these ruined houses had died, and that the one or two survivors had taken to flight. It would not be difficult to collect from the reports of district officials a number of such instances, but their collection could serve no useful purpose, as we should be still as much as ever in the dark as to the extent of the prevalence of the disease.

The provincial vital statistics throw no light on the subject. Taking the Kámrúp district as the best for our purpose, because *kála-azár* has only appeared within its limits during the last few years, one would expect to find the annual mortality for these years markedly increased; but, whatever be the extent of the fatality, it does not show itself in the returns. During 1889, the total mortality of the Kámrúp district *per mille* was 27·20, which is only 0·29 *per mille* in excess of the average death-rate of the preceding five years, an amount well within the ordinary limits of annual variation, as may be seen by an inspection of the annexed table, which gives the death-rate for each year from 1882 inclusive, during which period the conditions and efficiency of registration has been fairly uniform. The birth-rates are also given, as the proportion they bear to the death-rates gives one a rough test of the comparative efficiency of registration during the years in question :—

Mortality of the Kámrúp district from 1882—89.

Year.	Total death-rate per mille.	Total birth-rate per mille.
1889	27·20	21·68
1888	27·64	23·18
1887	22·83	24·27
1886	29·18	25·02
1885	27·59	26·00
1884	27·89	24·95
1883	25·60	22·66
1882	32·88	27·65

The great annual fluctuations, and their capricious relationship to the birth-rate indicate of course that registration
is as yet in its infancy in Assam, a fact which indeed is admitted, and adverted to, in each successive annual report. Still,
however defective registration may be, and however misleading as a measure of absolute mortality, the figures must have
some comparative value, and strongly suggest the conclusion
that the mortality from *kála-azár* has been somewhat exaggerated.

We hear nothing of *kála-azár* in the Kámrúp district up to
the year 1884. Then, creeping gradually up through the Goál-
para district, *kála-azár* first appeared in Kámrúp in 1885, and,
since then, has spread through the whole district, and is now
commencing to attack that of Nowgong. In spite of this, there is
no marked increase in the registered mortality. The year 1886
has, it is true, a mortality considerably higher than 1885, but
this is followed by a year of abnormally low mortality, in spite
of the fact that reports show that *kála-azár* was steadily on the
increase. The years 1888 and 1889 again return closely to
the average of preceding years.

In the Goálpára district, the death-rate has always been
higher than in Kámrúp, a fact accounted for, in successive
annual reports, as due to more efficient registration, but the
appearance of *kála-azár* is so nearly coincident with the introduction of an improved system of registration that the
figures are valueless for our purpose.

The fact appears to be that *kála-azár* enhances the sick rate more than of mortality. As to the number of sick, we have no direct means of judging, but it must be very large, as is shown by the serious decrease in the revenue of the affected districts, owing to a considerable area having fallen out of cultivation from want of hands to till it.   •

A man, so ill as to be unfit to work, is just as unable to cultivate the land as a dead man, so that no doubt incapacity as well as death accounts for a share of this abandonment of cultivation, and may explain why it is that the revenue returns are so much more affected than the vital statistics.   •

During the past year, in the district of Kámrúp alone, the falling off in the revenue attributed to *kála-azár*, amounted to Rs. 12,896, and as the separate holdings are as a rule very small, few paying more than a few rupees per annum, this must represent many hundreds of families either dispersed, or so diminished by sickness and death as to be unable to cultivate the land they had hitherto held. While, however, the increased mortality has not been sufficient, at any rate in Kámrúp, to make its mark in the very imperfect vital statistics of the whole district, there can be no doubt that, in individual villages the increase has been so enormous as to attain a truly pestilential character. The fact is that the distribution of the disease is extremely capricious, while the individual outbreaks are sharply circumscribed; one village being decimated, while another, close by it, remains unaffected, but, wherever the disease makes its appearance, in the course of a few months, large numbers of people are attacked, and the mortality of the place is enormously increased. This is illustrated by a table, prepared in 1884, by the Civil Surgeon, Dhubri, in which an approximation to the mortality from *kála-azár* is obtained by tabulating the recorded mortality from "fever" in the affected district. In the registers, all cases of *kála-azár* are of course included under this term, and it is well to remember that, in Indian vital statistics, the heading really stands for nearly all diseases except bowel-complaints, violent deaths, small-pox, and the small percentage of cases seen by medical officers. When tested by actual medical investigation, as I have seen done in the Central Provinces, the number actually due to malarial fever is found to form a comparatively insignificant percentage of the whole, but, as a general rule, in all parts of India, somewhat the larger half of all deaths recorded will be found to be included under this very elastic heading. Now the probable (not the recorded) mortality of the Goálpára district may be taken as nearly 50 *per annum per mille,* and we shall not be far wrong if we assume that 28 *per mille* of this total are cases which would ordinarily be returned under the head of "fever," so that any excess over this number in any given village may fairly be ascribed to the epidemic. In the above-mentioned table, reproduced below, Dr. Dobson gives the population and numbers of death from fever in 266 villages.

Name of village.	Census population.	Number of deaths from fever in 1883.	Deaths per thousand.	Name of village.	Census population.	Number of deaths from fever in 1883.	Deaths per thousand.
Rangjuli Outpost.				*Rangjuli Outpost.—(Concluded.)*			
Amjonga	583	20	34	Fakirpara	313	15	47
Pulana Chita	440	11	25	Khootabario	459	22	48
Kakomapura	188	11	58	Ambareobeemán	333	13	39
Rangpur	91	10	109	Khakapara	299	17	56
Kashumario	232	20	86	Masalam	455	16	35
Bangpura	257	15	58	Ghillabario	550	14	25
Shupuku	375	10	26	Shikajooho	1,123	78	69
Shilabania	137	10	73	Maniskpur	699	8	11
Dorakpura	139	16	115	Sachapance	528	9	17
Chormureo	180	22	122	Deulgooree	552	59	106
Allibario	376	24	63	Kharashinook	583	46	78
Kamarpotta	734	51	69	Chakbareo	...	10	...
Budanang	896	46	51	*Salmara Outpost.*			
Darrangeeri	3,397	284	83	Mandalgrám	741	135	182·1
Rangjooho	3,497	326	93	Ulortala	655	40	61·06
Rahamaheo	486	42	86	Kharaoram Tálábaree.	805	40	49·6
Bhomrapather	91	11	120				
Rangpather	111	11	99	Khamarshialmario	49	4	81 6
Kutakoothea	1,806	72	39	Dámrah	282	33	117·2
Dhikdhok	388	9	23	Hapangeereo	557	39	70·01
Deegboho	1,280	19	14	Hadeerampara	57	5	87·7
Pipmebareo	970	35	36	Rangchee	136	8	58·8
Maoang	1,302	63	48	Ghorapota	168	21	125
Atheabareo	1,243	45	36	Uporparah	189	7	37·03
Mámagáo	470	2	4	Habanipara Bongao	190	26	136·8
Ambook	598	27	45	Bamonpanie Khata.	184	9	40·9
Katalmooree	555	14	25	Khamarie	138	19	137·6
Gatiapara	498	25	50	Bambooputa	117	12	102·5
Arimari / Monbari Khamar.	286	5	17	Borapather	268	21	78·3
				Chemipeareo	219	50	228·3
Dhontolá / Gora Chatka	794	19	23	Bakharpara	66	6	90·9
				Dainrahat	345	9	26·08
Bamoonigao	400	15	37	Boralumpara	143	20	139·8
Tipnai	680	32	47	Hatibanda	314	4	2·11
Potpara	1,293	47	36	Changmareo	233	3	12 08

Name of village.	Census population.	Number of deaths from fever in 1883.	Deaths per thousand.	Name of village.	Census population.	Number of deaths from fever in 1883.	Deaths per thousand.
Salpara Outpost.—(Concluded.)				*Salmara Station.*—(Continued.)			
Bogalamarie, Khacrepara.	415	41	98·7	Shimlabari II. ...	117	1	8
Badershee	154	11	71·4	Rupaah	227	9	39
Kharapara	142	26	183·09	Uzarparah	49	9	183
Salpara	389	46	118·2	Kirtonpura	90	5	55
Falpanee	59	3	50·8	Bezupara	100	6	60
Guruah	565	9	15·9	Haldibari	125	6	48
Satapara	...	...	...	Kaltapara	54	10	185
Boramatiah	154	29	188·3	Narabhita	63	2	31
Hatimara	360	36	100	Kokila	1,095	18	16
Pepiraphar	61	17	278·6	Nomborpara	373	13	34
Pádfán Poozahoriparah.	753	55	73 04	Tainbulbari	59	1	16
Arrimari	265	10	37·7	Rangapanee	382	11	28
Ashu Churee	136	7	51·4	Lengtisinga	1,409	36	25
Chamgiapara	40	1	25	Noah Shasroh Bejoypur.	39	1	25
Shonan Doobi	179	51	284·9	Margao	32	1	31
Shiaine	293	5	17·06	Golapara	90	2	22
Kishnai hát	142	1	7·04	Tilpukhri	205	3	14
Kouripara	71	6	84·5	Anishury	111	9	81
Khokan	559	4	7·1	Peradhora	489	20	40
Chinapara	219	15	68·4	Dumuriah	776	10	12
Belpara	178	18	101·1	Khudro Narikola ...	229	4	17
Kasbpara	59	4	67·7	Jolakhara	197	10	50
Nokshar	82	10	121·9	Dholagao	844	21	24
Yera	228	46	201·7	Damorshurie	135	9	66
Karkalichachapani	257	10	38·9	Boragirigao	273	19	69
Barigram	166	32	192·7	Nodipori	272	5	18
Dámán	217	70	322·5	Arimara	131	10	76
Ning Yanee	359	13	36·2	Bora Narikobe	1,066	41	38
Nahadon	196	23	117·3	Rohumarie	1,805	74	41
Dhaigram	240	14	58·3	Sonamoobe	333	33	99
Khoruapura	28	4	142·8	Malaygorh	287	13	45
Kailumostra	87	11	126·4	Jhitkibario	194	1	5
Sardarpara	246	36	146·3	Pochamiah	651	21	32
Sianmari Jonaimari	239	38	158·9	Hooramara	338	5	14
Daboho	446	9	20·1	Charupooniah	205	16	78
Nananonubasa	279	33	118·2	Jaldhapara	64	7	109
Degohe	292	9	30·8	Chakro Bhum	38	1	26
Shaiknatu	429	19	44·2	Kákoijala	1,962	90	45
Ranasara	99	5	50·5	Kuchea Kata	496	26	52
Odeahpara	128	16	125	Shiporphila	792	29	36
				Sikkagao	520	42	80
Salmara Station.				Dolaigao	747	24	32
				Birjhara	1,701	50	29
Ghoramara	268	10	37	Boshongao	785	34	43
Shimlabari	52	4	76	Mulagao	881	24	27
Bara Ghoriah	327	8	24	Bongaigao	956	16	16
Chaklapara	140	1	7	Majgao	503	14	27
				Bamongao	273	19	69

Name of village.	Census population.	Number of deaths from fever in 1883.	Deaths per thousand.	Name of village.	Census population.	Number of deaths from fever in 1883.	Deaths per thousand.
Salmara Station.— (Continued.)				**Salmara Station.—** (Concluded.)			
Madaleemari	144	2	13	Singimara	667	25	37
Khongdon	236	21	89	Bhootkoori	274	3	10
Kali Doba	254	11	43	Sakh Chor	111	10	90
Chakepara	1,600	69	43	Shonkorkhola	1,144	45	39
Khokorpura	768	30	39	Shoznabhitu	291	11	38
Shalmara	1,116	47	42	Bashbari	1,351	72	53
Deohalty	1,411	42	29	Dewangao	523	53	101
Tulonga	390	13	33	Bashigao	1,057	23	21
Choutaki	541	18	33	Dhontola	282	14	49
Dompara	210	6	29	Barnipari	104	2	19
Balubario	168	14	83	Bazikpara	246	22	89
Chandaparah	56	3	53	Poschagao	882	24	27
Chukani	24	2	83	Paknabari	299	4	13
Mamarpara	120	28	233	Kaetpara	315	24	76
Pakhru Ghoori	107	1	9	Shakhotiegram	861	32	37
Jogughopa	729	34	46	Napatpara	215	1	4
Kakoitaree	666	28	42	Tanglamari	...	1	...
Udashirlitta	112	8	71	Sheejorgao	...	1	...
Chalondapura	1,569	65	41	Chabiechira	...	1	...
Katushbari	777	36	46				
Bagrabari	270	14	51	**Goálpára Station.**			
Udobi	154	4	26				
Boalmari	149	12	80	Garkuta	143	1	7
Kerkhabari	374	18	48	Falpanee	178	8	44
Khoragao	848	31	36	Kokonga	539	31	57
Alloknagar	109	15	137	Kukoria	348	14	40
Chappara	130	19	146	Karipara	1,202	44	36
Kaimario	60	2	33	Lalubarie	216	35	162
Balapara	340	11	32	Tongabari	1,241	1	8
Koriagao	119	1	8	Bhozmahi	433	8	18
Birpara	208	1	4	Thalpara	137	15	159
Amshooree	580	44	75	Bamunpara	...	...	...
Shooirmukha	91	1	11	Nowhata	191	1	5
Kheloorapara	52	1	19	Baraparamatia	202	1	4
Baripookhoori	433	51	117	Jugupara	79	7	88
Bongao	100	4	40	Manikpore	176	28	109
Neemagao	141	11	78	Dhandipara	97	7	72
Barakhata	519	29	55	Bhalakhamar	360	11	30
Beziemari	204	4	19	Khalmohora	117	26	222
Talshurie	255	13	50	Boramohara	485	39	80
Moobeegao	560	17	29	Bhakorbhita	310	3	9
Raghoonandanpur	856	39	45	Poncha Raton	112	1	9
Sal Bari	353	9	25	Makorie	181	2	11
Kasharpara	263	12	45	Khorboza	85	1	11
Bhishnoopur	835	44	52	Bhahekdubi	458	20	43
Borakhola	60	4	66	Doshakatol	713	10	14
Kumsakuta	78	1	12	Rohatie	1,102	49	44
Ghibishoorio	761	49	64	Garmari	78	3	38

Name of village.	Census population.	Number of deaths from fever in 1883.	Deaths per thousand.	Name of village.	Census population.	Number of deaths from fever in 1883.	Deaths per thousand.
Goálpára Station,—(Continued.)				*Goálpára Outpost.—*(Concluded.)			
Bhardobi	84	42	500	Mornai	836	81	96
Sheinlitola	548	3	5	Dalgao	636	21	33
Balipura ...•........	223	4	17	Bagooah	122	11	90
Shatorpara	349	2	5	Bakaitari	1,169	49	41
Takora	492	32	65	Badur Chor	624	24	38
Bhukpara	751	30	39	Hadagram	502	30	59
Dakoidol Terikoriahparah.	263	53	210	Mozkubi..............	307	11	35
				Rakhoshuhi	106	12	113
Doreemornai ...... {	694	...	...	Aoorkakuchee	556	14	25
Dobola .... {	964	61	30	Kokorin......	164	7	42
Gaobari .... {	303	...	......				

The total population of the villages included in this table is 117,912, and the number of recorded deaths 5,848, equal to 49·1 per mille. Subtracting from this the 28 *per mille* assumed as the normal proportion of deaths recorded as "fever," we are left with a mortality of 21·1 per thousand, equivalent to 2,488 deaths, due to some unusual cause, in a single year, and that in only a portion of the Goálpára district. As may be seen, in forming this estimate, all tendency to arrive at an exaggerated estimate has been avoided by taking the recorded mortality from "fevers" as being exact, while assuming the normal mortality of the district to be considerably above that recorded; but, in spite of this precaution, the rates from individual villages are something appalling, as will be seen from the selection from the preceding table given below :—

List of villages in Kála-azár stricken portions of the Goálpára district in which the mortality from " Fever" alone exceeded 200 per thousand in 1883.

Name of village.	Population.	Deaths from fever.	Mortality per thousand.
Champeani	219	50	228·3
Reperaghor	61	17	278·6

Name of villages.		Population.	Deaths from fever.	Mortality per thousand..
Shoran Dubi	...	179	51	284·9
Jira	...	228	46	201·7
Daman	...	217	70	822·5
Mamarpara	...	120	28	233
Khal Moharer	....	117	26	222
Bhardoki	...	84	42	500
Dakaidol (Jarporlahpara)	...	268	53	201
Averages	...	148·8	38·3	257

List of villages in which the mortality from "fever" alone exceeded 150 per thousand in 1888 in Goálpàra.

Name of village.		Population.	Deaths from fever.	Mortality per thousand.
Mandagrain	...	741	195	182·1
Kharapara	...	142	26	183·09
Boramatiah	...	154	29	188·3
Bangrain	...	166	·82	192·7
Senamari Jonaimari	...	239	38	158·9
Azarpara	...	49	9	183
Kallapara	...	54	10	185·2
Lalabario	...	216	35	162
Manikpur	...	176	28	159
Averages	...	215·2	38·0	176·5

There is no particular reason to believe that similar statistics for the present year would exhibit any great change from those obtained six years ago, but, at any rate, no improvement can be expected, as there cannot be the least doubt that the disease is now far more widely spread than it was in 1884, so that such a list would now probably exhibit an even larger proportion of villages with heavy mortalities. At any rate, no more recent statistics are available.

A glance at the above tables will show one notable point. This is that, with but one exception, the villages thus remarkable for so pestilential a mortality are of small size, and the worst instances occur in very small villages. On the assumption that the disease producing this mortality is communicated from man to man, this can be easily understood. A village, *e.g.*, such as Bhardoki, where half the entire original population of 84 souls died in this one year, would consist of but two or three families, in the Indian sense of the word; in other words, the inhabitants would be in close and constant association; and the high mortality, assuming the deeply-seated popular conviction of the communicability of the disease to be well founded, would be inevitable.

Thus, the average population of the villages having a mortality of over 200 per thousand is 148, that of the villages with mortalities between 150 and 200 per thousand is 215, while the average population of the whole number of villages, of all mortalities is 443. It seems to me that it is impossible to explain this fact, save on the assumption of communicability, for it is perfectly inexplicable that malarial poisoning, or any other non-communicable malady should thus single out the small villages for a disproportionately large mortality, scattered as they are, amongst the large ones, over all parts of the affected district; for the number of small villages exhibiting exceptionally high rates of mortality is too large to admit of the explanation of its being due to the fallacy of drawing statistical ratios from a small number of data. Thus, the deductions derivable from such statistics as are available all point to the conclusion that, whatever it may be, *kála-azár* is a disease which affects intensely scattered local centres. That while the mortality of such places is so serious as to cause a widely-spread and well justified panic, the total number affected, while sufficient to render unfit for work a number adequate to seriously affect the revenue of the affected districts, is yet insufficient to make any distinct mark on the vital statistics of large areas, even when they are as seriously affected as the Kámrúp district is at present. Finally, the distribution of the malady points strongly to the conclusion that *kála-azár* is a communicable disease.

IV.—FACTS RELATING TO THE SPREAD OF KÁLA-AZÁR.

In Appendix A to the Annual Sanitary Report of Assam for 1882, Dr. Clarke, the then Sanitary Commissioner, gives a short abstract of what was then known as to *kála-azár*. According to this report, the disease had attracted attention as far back as the year 1869, and appears to have been, up to the date of the report, confined to the Gáro Hills and those portions of the Goálpára district bordering on them. . During the next three years, it spread widely through the Goálpára district, affecting first those portions nearest the Gáro Hills, and gradually spreading till almost the whole district became dotted with affected villages. In 1884, the disease became so serious a matter, in this district, that Government organized special measures of medical relief, by starting dispensaries at several suitable centres, and employing a number of medical subordinates to travel about and visit the people at their houses. Accounts of these operations figure in several succeeding reports, · but they appear to have been attended with but little success. This was mainly owing to the difficulty of inducing the people to submit to European medical treatment, and their impatience for an immediate cure even when persuaded to give our methods a trial. They expected to be cured by a single dose of medicine, and seldom could be induced to take a second.· The disease being a very chronic one, no method of treatment - could be expected to yield any appreciable results at once, and I have found that the same difficulty exists at the present day. The treatment, however, was entirely devoted to combatting malaria, and hence, from the point of view of the disease developed in the prevent investigation, it is by no means surprising that success was so wanting, as it would only be in cases of true malarial cachexia, such as are always to be abundantly found in Assam, that success could be expected. In such cases, no doubt, considerable good was effected, but, for all practical purposes, the spread of the disease was in no way affected by the relief operations.

In the sanitary report of the Kámrúp district for 1886, we find Dr. Mullane noting the appearance of the disease in the western portion of his district, where it marches with the Goálpára district, and year by year we find the Civil Surgeons of Kámrúp speaking in stronger and stronger language of the ravages caused by the disease. In 1888 the disease had become most serious in Chaygaon, 30 miles from Gauháti, having taken, as is remarked by Mr. A. C. Campbell, the Deputy-Commissioner of the district, in his comments on Dr. Borah's annual report for 1888, four years to travel 36 miles, and during the latter part of this year, and during 1889, cases began to appear in Gauháti itself. At present it is undoubtedly increasing in the town, and there are numerous villages, close by it, which are as badly stricken as any of those in the Chaygaon district. It is now to be found well to the eastward of Gauháti, and is said to have invaded the portion of the Nowgong district contiguous to Kámrúp. On the northern bank of the river it has spread to Barpeta and Mangaldai, and is very severe in some of the villages just opposite Gauháti.

It is a noticeable fact that, once the disease has made its appearance in a district, it never leaves it, the weekly returns of the prevalence of epidemic disease showing it to be present to-day in every part of the country where it has been hitherto reported.

It will thus be seen that the progress of the disease is peculiar. It does not appear suddenly and inexplicably over a large area, like influenza or cholera : it does not spread rapidly from man to man like small-pox, and other specific fevers ; but it creeps slowly from village to village, holding what it has seized, but progressing so slowly that it has taken seven years to reach Gauháti from the Gáro Hills, a distance of not 100 miles. It must, I think, be admitted that these facts are in no way compatible with the theory of a malarial origin. Assam always has been extremely malarious ; but if, as has been suggested, *kála-azár* be due to waves of " periodic epidemic intensity " (of malaria), how is it that we hear of no such periods of intensity before the appearance of this disease ? and how is it that once it has appeared, the intensity remains intense, but ceases to be periodic ? Again, if *kála-azár* be but malarial cachexia intensified by the proximity of uncleared jungle, the habits of the people, and so forth, it is clearly incumbent on the advocates of this theory to show that these conditions and habits have been intensified in affected villages coincidently with the outbreak of the disease, but, in point of fact, no one pretends that any such change has taken place. Further, I would ask, how is it that it is common enough to find villages quite unaffected by it in the midst of villages badly stricken, which may be less than a mile distant ? And yet, that such is the fact, is a matter on which any one may

satisfy himself by visiting the affected districts. To take a con-
crete instance. About six miles from Gauháti is a village called
Paru, which is most seriously affected with *kála-azár*. The village
stands on alluvial ground, on the river bank, but the site is somewhat elevated above the general level of the plain by the detritus
washed down from the Kamakhia hill, the river bank being here
exceptionally high. To the east, the close but isolated
patch of jungle, clothing Kamakhia hill, reaches nearly down
to the village, but is separated from the habitations by a belt
of cultivation. On the north is the river Brahmaputra, and
the remaining sides look on to extensive open cultivated plains.
The houses are, for an Assamese village, closely placed, the
place boasting of two quite distinct streets, and there is comparatively little jungle within its precincts. The supply of drinking-
water is taken from the river, which is so absolutely at every
one's door that it involves less trouble to do so than to seek for
it elsewhere. The river here sweeps by the village in a strong
deep current, and there is no backwater to be found nearer
than Gauháti. On account of its proximity to Gauháti, I
visited the place many times, and have been into every corner
of the village, and am positive that there are no wells to be
found in it. Between the village and Kamakhia hill is a small,
very filthy *bhíl*. It is evidently used for watering cattle,
but the water is so filthy that I doubt if even an Assamese
could be persuaded to drink it ; and it is further from the houses
than the river is, so that it would involve additional trouble to
take drinking-water from it. It does not appear to contain
any fish. The neighbourhood of the houses is of course filthy
in the last degree, but in this matter it differs in no way
from every other Assamese village I have entered. The
other village, that of Maligaon, is almost three-quarters of a mile
from Paru, and is situated on the Trunk Road not far from
the spot where the path to Paru diverges. It is thus a long
distance from the river, and the ground on which it is placed lies
somewhat lower, but it is about the same distance from the
jungle of the Kamakhia hill. The hills to the south of the road
are nearer and shut it in more than Paru, so that there is quite
open country to the north and west only. The water-supply
is from two or three shallow *kutcha* wells, which contained, at
the time of my visit, only a little brown, evilly-smelling water.
It, too, has one or two pools, the counterparts of that at Paru.
In the matter of filth, it is neither better nor worse than the
other place, and, on the whole, it must be admitted that,
owing to its more open and slightly drier site, and its pos-

session of an excellent water-supply, Paru is by far the more desirable place of residence. Yet Paru, as we have seen, is being decimated by *kála azár*, while I could not find a single case of it at Maligaon, either described as such by the inhabitants, or in the shape of early cases diagnosable by anæmia, but not yet sufficiently affected to consider themselves ill. Why then should a wave of " periodic epidemic intensity " pass over the one place and not involve the other, so close by it, and so similarly situated ? Another illustration of the want of connection between the incidence of the epidemic and ordinary sanitary conditions is to be found in the water-supply. While on tour, I searched large numbers of specimens of drinking-water microscopically for *dochmius* embryos, but without success, and, after I reached Shillong, I determined to make a further attempt to find them, and accordingly, by the kind aid of Dr. Costello, the Sanitary Commissioner, obtained from the district medical officers a considerable number of specimens of drinking-water from infected localities. Having the conveniences now at hand, I supplemented the microscopical, by a rough chemical analysis. The results show the greatest variety, from really good drinking-water to the most horrible filth imaginable. In only one instance were anything like *dochmius* embryos discovered, and, for many reasons, even in this case, the exact character of the single nematode observed is very doubtful. The results of these examinations of drinking water are given in the subjoined table. The analysis of the Shillong pipe-water, which is really excellent, examined on the same plan, is given as a means of comparison :—

Results of analysis of Specimens of Water from

Name of place.	Smell.	Colour.	Reaction.	Chlorine as Nacl., grains per gallon.	Oxydiable matter as oxygen required to oxidize mgs. per litre.
Shillong pipe-water	*Nil.*	Clear and sparkling.	Neutral ...	1	0·5
Bázár well	*Nil.*	Good	Ditto ...	2·9	1·2
Label illegible (probably also from Bara Jalinga).	Foul	Fair	Acid	1·7	3·6
Nagadum tea-garden river-water...	Disagreeble.	Good after settling.	Neutral ...	1·6	0·7
Chunsali tea-garden well in coolie lines.	Very foul	Brown	Acid	7·1	4·8
Bara Jalinga tea-garden tank-water	Foul	Good.........	Slightly acid	1·7	3·1
Well at Kalapani, Salpara outpost.	Slight ...	Very brown	Ditto ...	2·6	2·8
Well, village Dighi, Darangiri jurisdiction, Rangauli outpost.	Little or more.	Good.........	Dittò ...	2·0	1·6
Well of Lakhipur, label nearly illegible.	*Nil.*	Fair	Alkaline ...	4·0	2·5
Bojni outpost, label otherwise illegible.	Foul	Good.........	Acid	1·2	8·0
Label quite illegible	„	„	„	1·2	3·0
Bijni outpost pipe-water	*Nil*	„	Neutral......	1·4	0·8
Darrangiri jurisdiction, label otherwise illegible.	*Nil*	„	Acid	1·6	0·8
Bijni outpost pond-water	Very foul.	Brownish...	Markedly acid.	1·6	12·0
Laokhoa *bhils*, Gauripore	Disagreeble.	Good.........	Acid	1·4	1·
Well Manicpore, Salpara outpost...	*Nil*	„	Neutral......	0·9	0·8·
Pond-water, Naipara, Goálpára police-station.	Slight ...	Fair	Acid	1·8	1·8
Khoody Mari *bhil*, Gauripur.........	Foul	„	„	1·4	4·0·

laces affected with Kála-azár and Beri-beri.

Total hardness, degrees, Clarke's scale.	Ammonia by Nestless test qualitative.	Microscopical appearances, &c.	Quality.
1·75	A slight trace ...	Scarcely any sediment, some algæ, copepods, and diatoms.	Excellent.
5	A trace ...	A very small amount of brownish yellow sediment, mainly mineral, no evidence of pollution.	Good.
1·5	Strong re-action	A moderately copious, brownish white, floculent sediment, consisting mainly of decomposed vegetable matter, cotton fibres, rice, husk, &c.	Very bad.
1·0	Ditto ..	Copious gray sediment, consisting mainly fine sand, diatoms, human hair, and some decomposed vegetable matter.	Bad.
3·8	Ditto ...	Sediment very copious, composed mainly of decomposed vegetable matter with mud, cotton fibres, epithelia, &c.	Very bad.
1·4	Very strong re-action.	A moderate amount of floculent sediment, consisting almost entirely of decomposed vegetable matter. copepods, diatomaceæ algæ, fibres, hair, epithelia, evidently much polluted.	Ditto.
1·4	Ditto ...	Small amount of red brown sediment, consisting almost entirely decomposed vegetable matter, copepods. No evidence of human pollution.	Bad.
2·4	A trace ...	Small sediment, mainly decomposed vegetable matter, copepods, a setigerous annelid, diatoms algæ. No evidences of human pollution.	Good.
4·9	Distinct reaction	Moderate sediment, mainly decomposed vegetable matter, cotton fibres and other evidences pollution, desmids, diatoms, infusorians.	Doubtful.
4·7	Very strong re action.	Scanty sediment, mainly organic, with a few living copepods.	Very bad.
4·9	Ditto ...	Scanty sediment, consisting of about equal parts sand and organic matter, a small cattle tick, copepods, &c.	Bad.
15·	A trace............	A very small sediment consisting almost entirely of desmids and diatoms.	Good.
2·4	Ditto..............	Scanty sediment mainly consisting of organic matter, with infusorians.	Ditto.
6·7	Marked re-action.	Considerable sediment, almost entirely organic remains of copepods; cotton fibres, and other evidences of human pollution.	Very bad.
1·9	Distinct	Small sediment, consisting almost entirely of filamentous algæ, living and decomposed ; cotton fibres, &c.	Doubtful.
1·4	A trace	Scanty sediment, mainly organic, with clayey matter, filamentous algæ, ostracods and copepods.	Good.
2·1	Distinct	Moderate sediment, almost entirely decomposed vegetable matter with fibres, hairs, and other evidence pollution.	Doubtful.
2·4	Very strong reaction.	Copious sediment consisting almost entirely of organic matter; filamentous algæ and diatoms extremely abundant.	Bad.

D

Results of Analyses of Specimens of Water from

Name of place.	Smell.	Colour.	Re-action.	Chlorine, Na Cl, grains per gallon.	Oxydiable matter as oxygen required to oxidize, mgs. per litre.
Backwater of Brahmaputra at Chowdoniah.	Faint......	Fair	Acid	1·0	0·9
Well, Pykan, Salpara outpost	Slight ...	,, ...	Alkaline ...	0·6	0·8
Banishpara well, Goálpára police hát	Disagreeable.	,, ...	Acid	4·0	0·7
Label illegible	Horribly foul.	Brownish...	,, ...	3·0	6·8
Well, Darrangiri, Ranguli outpost	Faint ...	Rather brown	......	1·6	1·1
Label partly illegible, Bijni outpost, well water.	Nil.	Good ...	Neutral ...	1·4	0·6
Label illegible............	Slight ...	Fair ...	Slightly acid	1·2	2·2
Ditto	Nil.	Good ...	Acid ...	1·0	2·4
Well, Chihari village, Ranguli outpost.	Slight ...	Fair ...	Slightly acid	1·6	1·7
Well, Kota Kuti village, Punjah outpost.	,, ...	Brownish ...	Markedly acid.	0·4	1·3
Lakhipur ("Radha para" well) ...	Hardly any	Good ...	Ditto ...	2·2	2·0
Chigknagoal	Slight ...	,, ...	Acid ...	2·45	1·0
Silchar Dispensary............	,, ...	Fair ...	Neutral ...	1·5	0·8
,, coolie depôt	Distinct ..	Brownish ...	Acid ...	3	1·3
Katalgouri Tea-garden *kacha* well..	Foul ...	Brown ...	,, ...	1·5	4·4
Kohlangaon............	,, ...	Fair ...	,, ...	5·0	...
Darangiri village	Nil.	,, ...	,, ...	0·5	...
Gabul ,, ...	Bad ...	,, ...	,, ...	0·3	...
Baringiri ,, ...	Nil.	,, ...	Acid ...	0·4	...

places affected with Kála-azár and Beri-beri.

Total hardness, degrees, Clarke's scale.	Ammonia by Nestless test qualitative.	Miscroscopical appearance, &c.	Quality.
3·0	Slight trace	Moderately abundant sediment consisting almost entirely of fine sand, with a little decomposed vegetable matter.	Good.
2·1	A trace............	Scanty sediment, mainly decomposed vegetable matter; diatoms, desmids, and other algæ very abundant.	Ditto.
12·2	Ditto	Scanty sediment mainly inorganic, with some decomposed vegetable matter.	Fair.
4·2	Strong re-action.	Copious sediment of tenacious clayey matter, containing much decomposed matter, vegetable and animal; desmids, copepods, fibres, and other evidences of pollution.	Very bad.
2·1	Very marked reaction.	Scanty sediment of inorganic matter, mixed with decomposed vegetable matter, copepods, amœbæ.	Doubtful.
10·3	A trace ..	Small amount sediment, mainly inorganic, with some decomposed vegetable matter, algæ, &c.	Good. .
1·6	Distinct re-action	Scanty sediment, mainly inorganic, infusorians ; an ovum (not of any known human parasite, but probably nematode.)	Doubtful.
0·7	Marked re-action	Scanty sediment, mainly decomposed vegetable matter, large numbers of infusoria and diatoms.	Bad.
2·8	Distinct reaction.	Soanty sediment, mostly inorganic, with a few copepods and diatoms.	Fair.
2·3	A trace ...	Scanty sediment, almost entirely decomposed vegetable matter, containing infusoria, with fibres, and other evidences of pollution.	Ditto.
2·4	Marked re-action	Much sediment; decomposed vegetable matter mixed with fine sandy particles, infusoria, cotton fibres, and evidences of pollution.	Doubtful.
1·6	A trace ...	Scanty flocculent sediment consisting entirely of zoogloea entangling grains of rather coarse sand. The zoogloea has probably developed since the collection of the specimen.	Good.
3·5	Ditto ...	Scanty sediment, mainly inorganic, diatoms, desmids, &c.	Fair.
5·8	Ditto ...	Considerable sediment, mainly decomposed vegetable matter.	Doubtful.
3·5	Distinct reaction.	Much sediment, clayey matter, with evidences of human pollution.	Very bad.
...	Marked ditto ...	Very scanty sediment, consisting almost entirely of fibres, particles of rice husk, &c., looking in fact as if a pinch of house dust had been added to the water.	
...	Strong ditto ...	Of the same character as above.	
...	Distinct ditto ...	Ditto ditto.	
...	Slight ditto ...	Ditto ditto but in addition a specimen of a *rhabditis*, possibly *dochmius*, was found.	

Result of Analyses of Specimens of Water from

Name of place.	Smell.	Colour.	Re-action.	Chlorine, Na Cl grains per gallon.	Oxydiable matter as oxygen required to oxidize ngs. per litre.
Nágaghooli tea-garden No. 2 coolie lines.	*Nil.*	Fair	Faintly acid.	3·3	0·8
Nágaghooli tea-garden No. 3 coolie lines.	Foul	,, ...	*Nil*	4·1	2·2
Hilika tea-garden, Bengali lines ...	*Nil*	Good.........	F a i n t l y acid.	2·1	0·4
,, hospital	,, ...	,, ...	Ditto	1·2	0·4
,, ,, jungle lines, 3...	Faint......	,, ...	Ditto	4·1	2·0
,, new lines, 3	Disagree-able	Fair	Acid	2·3	3·6
Talap tea-garden, Poolibari lines well.	*Nil*	Good.........	F a i n t l y acid.	1·4	0·8
,, ,, Dhangori lines 2nd well.	,, ...	,, ...	Ditto	0·9	0·4
,, ,, northern s i d e well, old lines.	,, ...	,, ...	*Nil*	5·3	0·4
,, ,, northern s i d e well, new lines.	,, ...	Fair	Disti n c t l y acid.	7·4	0·4
Talap new lines, southern side well	,, ...	Good	F a i n t l y acid.	3·7	0·4
Doom Doona, Muthik lines	,, ...	Deep brown	Acid	1·1	5·4
Bisakòpi tea-garden, Buch well No. 2 lines.	,, ...	Fair	F a i n t l y acid.	3·7	0·4
,, ,, Pipe-w e ll No 3 lines.	,, ...	Good	D i s tinctly acid.	1·7	0·7
,, ,, J u n g l e side well No. 6 lines.	,, ...	,, ...	*Nil*	3·1	1·0
,, ,, Big well No. 6 lines.	,, ...	Slight	Acid	0·9	0·4
,, ,, well No. 7 lines.	,, ...	Pale brown	*Nil*	1·0	1·0
,, ,, No. 9 lines.	,, ...	Slight ...	F a i n t l y acid.	1·4	0·4
,, ,, No. 2 lines.	,, ...	Fair ...	*Nil* ...	0·5	0·3
,, tea-house well ...	,, ...	Slight ...	Distin c t l y acid.	1·4	0·5

places affected with Kála-azár and Beri-beri.

Total hardness, degree, Clarke's test.	Ammonia by Nestless test qualitative.	Miscroscopical appearance, &c.	Quality.
5·1	Faint re-action ..	A copious brown sediment, consisting mainly of clayey matter, with some decomposed vegetable debris, infusoria, &c.	Fair.
3·8	Strong	A moderate amount of floculent sediment, almost purely organic and showing plain evidences of human pollution. Some large ova, possibly trematode.	Bad.
3·5	Trace	Scanty brown sediment, consisting almost entirely of decomposed vegetable matter.	Good.
3·5	Ditto	Sediment somewhat less copious, but otherwise of the same character as the preceding.	Ditto.
3·2	Strong	A moderately copious sediment, almost entirely decomposed vegetable matter, but showing some evidences of pollution.	Bad.
3·5	Distinct	Sediment less copious, but otherwise of the same character as the preceding ; monads.	Ditto.
3·8	*Nil.*	Scanty brown sediment, mostly decomposed vegetable matter, insect larvæ, monads, and infusorians.	Good.
2·8	*Nil.*	Very scanty sediment, mainly vegetable matter ; but fibres, &c.. of suspicious character.	Ditto.
6·3	Trace..............	A very scanty brown sediment, consisting almost entirely of decomposed vegetable matter.	Doubtful.
5·3	Ditto	A very scanty sediment, consisting almost entirely of fine sand, with a little decomposed vegetable matter.	Ditto.
4·9	Ditto	A very small sediment, mainly inorganic, with some decomposed vegetable matter.	Good.
3·8	Doubtful	Colour too brown for reliable NH_3 determination. A copious dark red peaty deposit consisting mainly of decomposed vegetable matter, and containing filamentous algæ, copepods, &c.	Bad.
6·3	Distinct	Copious orange-coloured sediment, consisting almost entirely of an oval bacterium.	Doubtful.
3·8	Marked............	A scanty dark brown sediment, consisting of little else than decomposed vegetable matter.	Fair. .
4·0	Slight	An exceptionally small sediment of the same character as the preceding.	Ditto.
3·5	*Nil.*	Scanty greenish sediment of the same general character as the preceding, but with some doubtful evidences of pollution.	Good.
3·5	Trace..............	A very small black sediment, consisting entirely of living and dead vegetable matter.	Ditto.
2·8	Trace ...;	Scanty brown sediment consisting of clayey matter, filamentous algæ, &c.	V e r y good.
3·1	Distinct ...	An exceptionally small, flocculent sediment, consisting entirely of living and dead vegetable matter.	Fair.
4·5	Ditto ...	Scanty sediment, consisting almost entirely of filamentous algæ.	Ditto.

Nor are these by any means isolated instances. Far from being so, cases such as that of Paru and Maligaon may be described as typical of the distribution of the disease, especially in the Chaygaon district, where the villages, though small, are very numerous, and placed at small distances from each other. In the earlier reports on the subject, and in support of a malarial origin for the disease, much stress is laid on the assumption that *kála-azár* is necessarily a terai disease ; and, no doubt, as long as the disease remained confined to the Gáro terai, this was a fact, but the spread of the disease shows that the connection, then so confidently asserted, was quite casual, being merely due to the circumstance of the disease, at that time, not having spread beyond the terai regions. Now, however, the disease is common in portions of the Province which are in no sense terai, and as far distant from the hills as any portion of Assam can be. For example, the Chaygaon thána is now the portion of Kámrúp most seriously affected, and the disease is as rife there as anywhere in the Province, so much so, indeed, that it was at once suggested as the most favourable centre for my enquiries; yet this very region was, previously to the appearance of *kála-azár*, considered the most salubrious portion of the district. Commenting on the comparative immunity from cholera enjoyed by the thána during the year 1882, Dr. Russell, the then Civil Surgeon, remarks :—

" Its natural features should make Chaygaon thána the healthiest region of the district. Its southern portion consists of hills and highlands, timber covered, and well watered by hill streams (branches of the Singra and Kulsi rivers). Its northern half, though lying low, and containing a large extent of swamp, yet is well tapped by watercourses, and consequently fairly well drained. It is rarely subject to inundations, and, if these occur, they quickly subside."

Nor is Dr. Russell alone in his opinion, for Dr. Mullane, in his annual report for 1884, remarks in much the same terms on the comparatively greater healthiness of the Chaygaon thána over the rest of the Kámrúp district. Added to this, by far the greater part of the population live far from hilly or terai portion of the thána, and are so closely settled that the land is almost continuously under cultivation, and exceptionally open and free from jungle. If malaria be the source of-*kála-azár*, how is it that a district possessing such comparatively favourable features should be now devastated by the latter disease, while many truly terai portions of Kámrúp remain untouched. It will be thus seen that the facts relating to the spread and distribution of the disease are entirely at variance with a malarial origin.

In connection with the spread of the disease, another point remains to be considered, and this is the strong popular belief in its contagiousness. This belief is founded on the undisputed fact that, when once a case of *kála-azár* has appeared in a village, others soon appear, or, in other words, that the disease appears to spread. Moreover, this tendency is especially often noticed in members of the same household, who, one after another, contract the disease until occasionally all may fall victims to it. So far as I know, no one has ever disputed that such are the facts of the case, and during my enquiries I found that if one person in a family was affected, it was common to find that many other members would be found in various stages of the disease ; while other large families remained quite free from the disease, though living in a badly-stricken village. Now, these facts certainly point strongly to contagion. They may, or may not, be capable of some other explanation, but in the face of their admitted genuineness, it appears to me quite unjustifiable to state, as is done in so many reports, "There is not a particle of evidence in favour of contagion." Nearly all the reports I have read confidently assert that there is no evidence to support the popular view ; but it is evident, on reading them, that the conclusion has been based rather upon a systematic ignoring the facts of the case than upon a patient consideration of them. Dr. P. M. Gupta, for example, in his report on the Gáro Hills for the year 1888, says :—

"The belief that it is contagious is founded in the fact that when one is seized with an attack of this fever in a village, *there are others soon to follow*" (the italics are mine); "but this is no doubt due to the same climatic influences under which they all live ; and that the fever ceases to be less virulent, and less liable to attack others when the inhabitants shift from one village to another" (as they not unfrequently do when this disease is very bad amongst them), "only goes to prove my contention" (of a malarial origin) "beyond doubt."

I select this report because it contains, in a few words, the leading arguments, over and over again repeated, in other reports, as to the malarial origin of *kála-azár*.

The second sentence is somewhat involved, but I take the writer to mean that removal from an infected site is efficacious in checking the spread of the disease.

It is perfectly obvious that the facts, which are admitted by the writer, really warrant quite the opposite conclusion.

It is perfectly clear that, from the nature of the case, in all instances of communication of disease from man, to man the subjects must be under the " same climatic influences," and that the argument might be applied with equal. cogency to small-pox, or any other admittedly infectious disorder. If, moreover, these outbreaks of *kála-azár* in villages be due to climatic in-fluences, it is clear that the outbreak should attack a large number of people simultaneously, or nearly so, whereas, instead of this, we find that the disease always attacks but few of the inhabitants of a village at first, and then spreads with excep-tional slowness. Accompanied, as it is among the Gáros, by desertion of the sick, it is obvious that migration is a most efficient measure in the case of any communicable disease, whereas the breaking up of new ground would, as is well known, tend to intensify malaria instead of checking its ravages, so much so, indeed, that other writers have, on this very ground, advanced the migratory habits of the Gáros as one of the causes of *kála-azár*, so that while one observer points out that migration checks the disease, another gives it as one of its causes, both, in different ways, twisting the facts to support a connection between *kalá-azár* and malaria.

If analyzed, all the arguments in favour of malarial origin amount to this. *Kála-azár* closely resembles malarial cachexia. Therefore they are identical. Malaria is not contagious. Therefore all facts that point to contagion must have some other explanation; which, as will be seen, is but arguing in a circle.

It is never safe to entirely ignore a widely-spread popular conviction of this kind, and I know of no instance of a disease popularly believed to be communicable ultimately proving to be otherwise, for the popular mind is much more apt to overlook contagion than to believe in it. Ten years ago, the medical profession were practically unanimous on the belief that tuberculosis was never contagious, and this in spite of a strong popular impression to the contrary. Yet, quite recently, the communicability of tubercular phthisis has become an established fact, and the methods and vehicles of communication have been clearly made out. As will be seen, when the method of transmission of anchylostomiasis is dis-cussed, the fact that migration accompanied with desertion of the sick is efficacious in checking *kála-azár* entirely fits in with the facts relating to the method of communication of this parasitic disease, and is a strong argument in favour of their identity. Indeed, if the arrangements for excluding the sick could be made thorough, it would be absolutely effectual, and the reason why *kála-azár* often springs up again amongst the migrated community is because the advance of the diseases is so insidious, that they necessarily included among their number many infected, but unsuspected persons.

Another point in connection with the spread of *kála-azár* is that there is no recorded or even hearsay instance of an European becoming infected. Europeans, of course, suffer much less than natives from malarial cachexia, but they nevertheless enjoy no complete immunity from it. The number of Europeans living in the affected districts is, it is true, but small, but, in spite of their small number, their escape would be indeed remarkable if malarial influences were abroad sufficient to account for the terrible ravages of *kála-azár*. For these Europeans were certainly exposed to the same climatic influences, and the conclusion seems inevitable that, as their climatic surroundings were the same, the source of their immunity is to be found in a difference of habits, and, be it observed, that it is only among such Europeans whose habits assimilate them to the semi-civilized inhabitants of Assam that anchylostomiasis has ever spread. It is only among the lowest class of miners and brick-makers that the disease has been ever known in Europe, never among the class who migrate to Assam to serve as officials, tea-planters, and missionaries.

V.—The Symptoms of "Kála-Azár" as illustrated by Cases.

What then is *kála-azár?*

The best method of answering this question will be to give the notes of a number of cases, as described by different observers, commenting on whatever may appear most remarkable in each or all.

I will first quote some cases described by Dr. Dobson, Civil Surgeon of Goálpára, in a special report on *kála-azár*, furnished by him, when the disease first reached an epidemic importance in that district. These ‘cases are convenient for our purpose, because, while occupying but little space, whatever is recorded is so much to the point, that it is generally easy to make a diagnosis of the true nature of each case, and they illustrate, in a most typical manner, the sort of cases that are brought to one, when visiting a *kála-azár-*stricken village.

That Dr. Dobson did not then recognise the greater proportion of them as instances of anchylostomiasis is due to the fact that the disease is one which is never encountered in English medical practice, and that for this reason we are at the best familiar with it only by occasional extracts from the foreign medical press. At the date of Dr. Dobson's report, the connection between the anæmia of coolies and anchylostomiasis was not even suspected. I had the pleasure of many discussions with Dr. Dobson during my stay at Dhubri in March last, and, since then, he has made renewed observations on the subject, and appears to have come to much the same conclusion that I have, for while "opposed to admitting all *kála-azár* is anchylostomiasis," in his remarks on the Damra Charitable Dispensary he says "I have no doubt many of the so-called cases of *kála-azár* are cases of anchylostomiasis;" and notes, too, that malarial cachexia and anchylostomiasis appear to complicate each other in Damra, which is entirely in accordance with my own observations elsewhere. This, it appears to me, is as near a coincidence of opinion as can be expected so shortly after the promulgation of my views in the preliminary note appended to last year's Sanitary Report, and I feel sure that a few more visits to *kála-azár-*stricken districts will bring Dr. Dobson's views into entire accordance with my own.

That people brought low by anchylostomiasis should more

readily become the subjects of malarial poisoning is only natural, and a very large proportion of cases undoubtedly do present
this complication; but, apart from some amount of increase on
this score, I see no reason for believing that malarial cachexia
is commoner now than it always has been. From first to last,
as much before *kála-azár* was heard of as after, the ravages of
malarial poisoning form a prominent subject of annual and
other reports. Indeed, the malariousness of Assam is notorious
now, and always has been, so that it would be indeed astonishing
if *kála-azár* cases failed to present this complication.

Case No. 1 "was that of a Gáro woman, who had been
ill for over a year. It had begun with strong fever. She had
tried village medicines, which consisted of leaves mixed with
mud; but as she got worse and emaciated, she was told she had
"*kála-azár*" and nothing further was done for her. When seen
by me, she had a temperature 103° F.; she had a troublesome
cough, with little or no expectoration, dullness over both lungs,
especially about their apices, and all the physical signs of phthisis were present. Her pulse 65, and very weak; her tongue pale,
conjunctivæ deadly white; eyes far sunk in their sockets; cheek-
bones prominent, cheeks hollow. Appetite almost *nil*. She was
so weak that she could not sit up, and, when raised, we had to
put her down. Her spleen and liver were not enlarged: if anything, they may have been slightly atrophied. She was passing
her stools involuntarily, and had done so for four or five days.
There was no dropsy or œdema of any kind present, nor could
I learn that she ever had either. Such then was a case of *kála-
azár*, but which was in reality a typical case of phthisis. The
woman's complexion was of the ordinary brownish colour. No
one would come near her, she was alone in a miserable hut,
which was tumbling down. Her food and drink were handed
in at the door, none of the villagers would touch her: even her
husband had taken another house in the village, and the
children (varying in ages from three to seven) called out to the
Hospital-Assistant and myself, and warned us not to touch her:
when they found we had, they immediately called out ' you
will get *kála-azár*.'

"This is the popular idea among the Gáros, and it is one that
is early impressed on the young mind. The odour within this
poor woman's house was very trying. It was difficult to
get even these few particulars about her illness, as every one
shunned her: even these particulars, such as they are, must be
received with caution."

In this case the most prominent symptoms, at the time of examination, were those of lung mischief. There can, however, be little doubt that the case was primarily one of anchylostomiasis, low broncho-pneumonia, such as exists in one form of phthisis, being one of the commonest terminations of the disease, and actual phthisis a common complication.* It is to be observed that there are no signs whatever of malarial cachexia, neither spleen nor liver being enlarged.

Case No. 2 " was that of a very healthy-looking girl, aged nine years, who was brought to me by her grandfather saying she had *kála azár*. It seems she had fever a year ago, which lasted for two or three months steadily, coming on at evening usually. From the description given of it, it must have been an ordinary case of quotidian ague. For the last two months she had no more fever. Three months ago, for about ten days, but at intervals of two or three days, her nose bled, but to no great extent. At this time also she had occasional attacks of diarrhœa. At present she is very well nourished, with a good appetite, a clean tongue, no pains, no pigmentary deposits of any kind ; her spleen extends down to her groin, and her liver is also enlarged. She is tolerably dark in complexion. The grandfather is aware of her enlarged spleen, and imagines she had *kála azár*, but does not think she has become blacker than usual."

This is evidently a case of ordinary malarial fever of a mild type. Observe that, in spite of very large spleen and enlarged liver, her general health is comparatively little affected, as so often happens in such cases, when tending to cure.

Case No. 3, " a Rabha man; fever for four months, which comes on every third or fourth day. He had no fever when I saw him. He is well nourished, but slightly anæmic. His appetite is good. Spleen is slightly enlarged. He is darker-skinned than most of his neighbours, but if anything slightly clearer than his own brother, who is in good health. He has been told he has *kála-azár*, but does not believe it, as he does not think he is likely to die for some time to come. He has never had diarrhœa or œdema of any part of his body."

Probably a case of early anchylostomiasis complicated with malaria.

Cases Nos. 4, 5, and 6 " are those of a Kachári man and his two sons, aged five and seven years. They have all

---

Vide Dr. Kinsey's pamphlet, already quoted, pages 52-53.

been ailing for four months only. A son, aged about four years, died six months ago of the same disease. Their illnesses began with strong fever, which still attacks them every now and again. [The fever is intermittent from the description given]. These three unfortunates are at present very emaciated and very anœmic. The father has œdema of both feet (very marked) and puffiness about his eyes. The youngest boy has also œdema of both feet and a collection of fluid in his abdomen and scrotum. The abdomen is so much distended that he is scarcely able to stand. The superficial veins of his abdomen are markedly distended. Both boys have very large spleens. The father's spleen is not enlarged. The livers in all cases are normal. The father alone has suffered from diarrhœa at times during his illness. At present the bowels of all three are irregular, and they all complain of great weakness and loss of appetite. They have noticed nothing abnormal about their urine, which seems right in colour and quantity. The man has lost all virile power for over two months. I failed to find any pigmentary deposits about this family, in fact, they are all of a light brown tint. They have never had any bleeding from the nose. The father and youngest boy are scorbutic. Their food consists of *dhal* and rice, and, from what I could gather, very little of it. The hut they occupied was a miserable one, low and dark, and thickly surrounded with jungle and rank vegetation. This family is in very needy circumstances."

These are all three typical cases of anchylostomiasis in an advanced condition. In the case of the two children, it is complicated with malaria, but the father is free from it. Here, too, is an instance of the way in which the disease spreads through a family.

Case No. 7 "is that of a Gáro woman, said to have had fever for the last four or seven months every now and again. She is slightly emaciated, but very anæmic. Her tongue is pale and flabby, appetite pretty good. Has never had diarrhœa, œdema, or bleeding from the nose. Her spleen is not enlarged. Her great distress seems to be due to an intensely severe pain over the left half of her head. This has troubled her for six weeks, and is scarcely ever absent. She is at present suckling an infant seven months old."

A typical case of anchylostomiasis moderately advanced, and without a sign of malarial complication.

Case No. 8 "is that of the Agia school-master. He has had fever (intermittent) almost daily for the last four months. He frequently feels feverish and out of sorts, although no fever is actually present. When I saw him, his temperature was 100·2F., pulse normal, never had diarrhœa, is very scorbutic, his gums were bleeding. Bled from the nose fifteen days ago, and again after three days : on both occasions the bleeding was profuse. His spleen was normal. His appetite bad, his bowels somewhat constipated. His conjunctivæ were very white. He is rather a dark-skinned individual."

Clearly a case of anchylostomiasis, again without any signs of malarial cachexia.

Case No. 9 "is that of a girl aged 14 years, who had been ill for ten months or more with fever, which she gets rid of for a time only, as she suffers from it off and on. She had diarrhœa, but no dropsy or epistaxis. She is slightly scorbutic. Her spleen is much enlarged. Appetite good. Bowels regular for some time past. Tongue clean. Her complexion is light brown, and she is intelligent and cheerful. She had attended the dispensary two mornings, with long intervals between the two visits, but, as she did not improve, the father informed me there was no necessity for her doing so any more. I attempted to show him the absurdity of his ideas, but he held to his own opinion on the subject."

An ordinary malarial case, but which hardly can be said to have reached a cachectic condition.

Case No. 10 "was that of a man, aged about 23, who had suffered from fever for the past eighteen months. He does not get it near so frequently of late, nor is the fever so severe as it used to be. He is weakly. His gums have bled some months ago, but he has no scorbutic symptoms at present. He has a fair skin, a huge liver and spleen, and œdema of both feet. His pulse is slow and weak. He is anæmic. His appetite is very good. He usually eats *dhál* and rice, and only rarely does he eat fish or fresh vegetables. He will not go to dispensary. He is at present being treated by a 'Kobiraj.' The treatment consists in keeping a spot on the crown of his head (about the size of a rupee) shaved, and into it daily is rubbed some preparation, the object being ' to cool the blood.' The application does not burn or smart in the least."

Also a malarial case, but which is in all probability complicated with anchylostomiasis.

Case No. 11. "was that of a young man aged 23, who was greatly emaciated. He had suffered from fever for several years, but only during the last eight months has become very ill. His spleen is enormous, his abdomen is greatly distended, both by his spleen and dropsical fluid. Very marked dropsy of both lower limbs as far up as the knees. The superficial veins of his abdomen are very prominent. He has frequently, at different times of his illness, suffered from diarrhœa, but it has clung to him with great pertinacity for the last six weeks. His urine is of a reddish brown colour, and he passes the usual quantity. He has never bled from the nose. He is scorbutic, and his gums have bled at times. On his trunk are purpuric spots and extravasations of blood. The mucous lining of his mouth shows a black pigmentary deposit. He is very anæmic, has no appetite. Now passes his stools involuntarily, Is not able to stand. His complexion is of a light brown colour. His brother says he has become black, whereas, in point of comparison, I remarked he is very fair for a native, and much fairer than this own brother of his."

A case of anchylostomiasis in its last stage, complicated with malaria and scurvy.

Case No. 12 "was that of a boy aged five years, who has been ailing for eighteen months. It began with fever, but after about four months the fever left. He is now in the same state very nearly as when I saw him on the 8th December last, the only change being a slightly greater enlargement of his spleen, which is truly enormous. His liver is also enlarged. His conjunctivæ are yellow. His abdomen is so greatly enlarged owing to the state of his viscera and dropsical effusion, that the poor little fellow, on being made to stand began to cry, lest he should fall down, as he was not able to support himself. When allowed to be seated on the ground, he was perfectly happy. The superficial veins of his abdomen were distended and prominent. There was no œdema of legs or feet. There was great emaciation. Complexion was of a rich brown colour."

A case of pure malarial cachexia. Observe how, in this case, ascites precedes œdema of the lower extremities. In anchylostomiasis œdema usually precedes ascites. In malarial cachexia of course, œdema often appears later on, but naturally ascites appears first, due, as it mainly is, to mechanical obstruction to the portal circulation by the pressure of the enlarged spleen and liver on the vessels. In anchylostomiasis, on the other hand, the dropsy is of hæmic origin, and naturally commences in the parts most distant from the heart.

Case No. 13 " is that of a policeman whose home is in North Salmara. His present illness dates four months back, but there is an old history of fever. He says he is unfit to do his work. He is emaciated and has an enlarged spleen ; occasionally suffers from diarrhœa, but has never had dropsy or bleeding from the nose. The least amount of work quite knocks him up. He is anæmic, but not scorbutic."

A case probably of anchylostomiasis and malaria, but it is difficult to say which factor predominates.

Case No. 14 " is that of a Rabha man, age not known, who has been ailing for fourteen months. His illness began with a strong fever (intermittent) which lasted for three months. At the end of this time he had diarrhœa for ten days. At present he complains of having a great disinclination to do work of any kind, and, should he do any work, he finds a trifling amount quite fatigues him. His spleen is slightly enlarged, and his conjunctivæ are very white. Pulse 70, but not strong."

This, again, is clearly a case of anchylostomiasis in a moderately early stage, with trivial malarial complication. This man's sister was also ill, but Dr. Dobson was not allowed to make any proper examination of her case.

Case No. 15 " is that of a Rabha man aged about 30 years. He was somewhat fairer in colour than his neighbours, and was suffering from œdema of both feet and seemed very weak. I was told by his friends that he had been ailing for a long time and that he had *kála-azár*. This man at the time of my visit was busy making a sacrifice."

This too is pretty certainly a case of anchylostomiasis, though the data are rather scanty.

Case No. 16 " is that of a boy aged 12 years. He has had fever for over a year. Fever present when I saw him. Very anæmic, liver and spleen both enlarged, the latter very much so. Conjunctivæ very yellow. Never had diarrhœa. No other symptoms noticeable ; complexion is of the usual brown colour. The peculiarity about this case is, he is said to have *kála-azár* and yet reputed to have got a yellower colour than he was when in health. His father asserts ' he has not got blacker, but fairer.' "

Appears to be a case of ordinary malarial cachexia, perhaps complicated with anchylostomiasis.

Case No. 17 "is that of a Rabha man aged 40, although, according to his own statement, he is not more than 20. He has suffered from fever for ten days. It is intermittent in type. He is altogether a debilitated subject and anæmic, his chief complaints being weakness and total loss of virile power, during the last three days only. He further stated this is usual in *kála-azár*, and that when the women get *kála-azár* their menses cease."

There is nothing here to account for the anæmia, but anchylostomiasis especially as the condition of the spleen is not stated, but enlargement cannot have been very marked, or it would have been noticed, as in the other cases.

Case No. 18 "is that of a Rabha man who has had fever for over two years. For the first four months he suffered most, and then he found he had a large spleen. The fever still attacks him occasionally. He now suffers from great weakness, and is quite unable to move beyond his door. He is extremely emaciated, his conjunctivæ are very anæmic. He has suffered from diarrhœa for the past two months. Never bled from nose. Is scorbutic. Belly greatly distended with fluid,the superficial veins of abdomen are prominent, spleen enormous. Liver seems normal. Great œdema of both feet. Puffiness about the face and eyes. Complexion light brown, but on the mucous lining of his cheeks there is a fair amount of bluish black deposit, which is very distinct. This man has two children aged about four and two and a half years of age. Both these youngsters have large spleen, and are thought to be healthy. The father has lost all virile power for the past four months only."

A case of anchylostomiasis in the last stage, the œdema affecting even the face. There is malarial and scorbutic complication.

Case No. 19 "is that of a chaukidar in charge of the Public Works Department rest-house. He has had intermittent fever daily for the last three months. He has become very thin and weak, and has no desire to do anything. He is very anæmic. His spleen is not yet enlarged. His appetite is very poor. He says he has '*kála-azár*.' He is of a light brown colour. He objects to take our drugs. I saw this poor fellow again some days ago, and he then had strong fever, 103°F."

A case of anchylostomiasis, with an intercurrent attack of malarial fever, but in the absence of any signs or history of chronic malarial disease, he certainly cannot be said to be a subject of malarial cachexia.

Case No. 20 "was a child eight years old who had fever two years ago, which lasted for several months, and which left the child very weak and emaciated and with a large spleen. The weakness is daily getting worse, and a few days at most will see the end. No further symptoms discernible, nor could I obtain any further history relative to this case."

This, on the other hand, is probably an ordinary malarial case.

While on this tour, Dr. Dobson, apparently mistrusting the sufficiency of malarial poisoning to account for what he had seen, examined 42 children to all appearance quite healthy and reputed to be so, and found that 21, or half the number, had enlarged spleen, the organ in some cases being very large. This accords with my own experience in the Chaygaon district, where I examined the spleens of a large number of people reputedly healthy. The proportion was about the same, but unfortunately I have lost the slip on which the notes were made. If then half the reputedly healthy population show more or less splenic enlargement, is it surprising that it should be found also in *kála-azár*? The reverse would indeed be astonishing, and in the face of such a fact it is obviously absurd to attach any pathognomonic importance to this symptom in connection with the etiology of *kála-azár*.

The next cases are from Mr. McNaught's series, but, owing to the detailed mass of daily notes, it is impossible to insert them in full, so that an abstract only is given:—

Case No. 21, Mantang, cultivator, age about 30, living at Kalputa in the Gáro Hills. Father died when he was a child, mother about a month ago. Is married, and has two children, said to be healthy. His village consists of about 16 houses, all inhabited by Gáros. A large number of people had died of *kála-azár* in neighbouring villages, but his village was not attacked till about a year ago. The disease was believed to have been brought from Khilonbari, which is within sight of Kalputa. A man called Thodang was the first person attacked, and, though ill, he still lives. This man's grown-up niece, and some of his children have since become affected. After this, a man called Melcha took the disease. Then Illal's household became affected, and four of his people are still ill. Two men died of the disease in separate houses, but the family of one of them were not affected. This illness commenced about a month ago with feverish symptoms and diarrhœa, then his spleen became enlarged and he began to lose flesh.

Present condition is weak, thin, and anæmic. Has slight anasarca, appetite bad, slight looseness of the bowels; spleen enlarged to 10 fingers below the ribs. Abdominal pain.

The man remained 3½ months in hospital, and, in spite of continuous anti-malarial treatment, grew steadily worse, had repeated bowel attacks, and died from an attack of pneumonia.

He is recorded as having fever pretty regularly at some portion of the 24 hours. In spite of this, his temperature, with one or two slight rises, remained subnormal during the whole of his stay in hospital, often not reaching the normal for a week together. During the last fatal attack of intercurrent pneumonia, it ran up to 103°, and remained above normal for some days. It had, however, been again subnormal (95·8°.) for some days before his death. The Hospital-Assistant's puzzlement at having to do with a case of fever, which the thermometer persistently belied, is betrayed by such phrases as "No fever now, but had fever during the night," never apparently suspecting that the man had no fever at all, and that his saying he had had it, merely meant that he felt ill, and that, like most sick people, he felt worse at certain periods of the 24 hours than at others. Anti-malarial treatment had a fair trial in this case. Quinine in 16 grains doses, arsenic, iron, and iodine were all tried, and persevered in; but, in spite of the fact that the man's case was at first by no means serious, and had only lasted a month, under this treatment the man steadily grew worse instead of better. Considering that the only symptom of malarial poisoning in this case was a moderately enlarged spleen, such as is found in immense numbers of Indians, to all appearance enjoying very fair health, this result is by no means surprising.

Post-mortem examination twelve hours after death.— Body emaciated, not darker than usual. Feet œdematous. Section of tissues anæmic.

Thorax left plural cavity contained fluid of a slightly turbid character, pleura congested and rough. Posterior part left lung in a state of gray hepatization, no fluid in right pleural cavity.

Heart contained P. M. clots ; structurally healthy.

Abdomen contains a moderate amount of pale serum.

Liver of normal size, not congested, of firm consistence, "otherwise of normal appearance." Spleen enlarged to three times normal size, lobulated, section firm, capsule thick and corrugated, no infarcts.

Kidneys—right congested, medullary portion slightly granular ; left of similar appearance. Suprarenal capsules appear normal.

Stomach contains thick white mucous adhering to mucous membrane, *the surface of which is brownish and corrugated*, but not ulcerated ; mesenteric glands of normal size ; section anæmic.

Intestines—*the duodenum and about 2 feet of the jejunum were of somewhat slaty colour, which became lessened gradually in the lower parts.*—Lower down the bowels were considerably thinned with patches of dotted congestion, sometimes extending through Peyer's patches in streaks, but not over the whole surface, but no ulceration, and there were other Peyer's patches quite healthy.

Here we have described a typical case of anchylostomiasis alike in symptoms, progress, and termination. *Postmortem*—the appearances found after death in such cases are all described, except the worms, which, as I well know, are much more easily overlooked than found, even when one is actually searching for them ; but the condition of the stomach, and of the duodenum and upper part of the jejunum are most typical, and a similar combination will hardly be found in any other disease than anchylostomiasis.

The thinning of those parts of the intestine not the seat of active irritation from the parasite, is of course only a part of the general emaciation, while the reddish streaks found were probably only patches of hypostatic congestion.

Case No. 22, Khejan, a Gáro cultivator, Æt. about 15 years, came from Kerapara in the Gáro Hills. "The usual history" of illness for some time, with enlarged spleen, fever, gradual wasting, and inability to work.

Present condition is anæmic, thin, and weak, but can still get about. Gets fever irregularly. Bowels regular, chest healthy, pulse quick, tongue pale, appetite fair, spleen slightly enlarged. None of his relatives have had *kála-azár*.

The after history of this case closely resembles the preceding. He was submitted to anti-malarial treatment, under which the spleen diminished in size ; but, in spite of this, he became steadily worse, and died from a violent intercurrent attack of diarrhœa, after a stay in hospital of a little less than three months.

A temperature of 101·8° is recorded once only, and with this single exception, there was never any pyrexia during his whole stay in hospital, and, as in the preceding case, the temperature is often sub-normal.

The *post-mortem* examination showed the death to be immediately due to enteritis. The same peculiar appearances are recorded in the duodenum, and he also describes a large number of "*deposits of pigment of an inky hue varying in size from that of a millet-seed to a split pea, many of them being quite firm to the feel.*"

These are of course the well-known "bites" made by the parasite, and consist of extravasations of blood into the mucous and submucous tissues. It is instructive to note that, whatever malarial element may have at first been present in this case, it was in a fair way to cure at the time of his death.

Case No. 23, Nausi, a Gáro cultivator's wife, aged 26. *Kalá-azár* had appeared in her village two years ago. She has been ailing for a year, her child took the disease and died, and her husband ran away from her, from fear of contracting it. Her illness commenced with pain in the head and discharge of blood from the bowels. She stated that her skin had become darker. Her appetite is bad, she complains greatly of headache. Her spleen is somewhat enlarged. She is thin and weak, face and eyelids puffy, conjunctivæ of a "muddy" colour, tongue pale and flabby. Base of lungs œdematous.

This woman remained nearly six months in hospital, during which time she had repeated severe attacks of diarhœa, the bowels being loose the greater part of the time. She grew decidedly better, the diarrhœa ceased, and she was discharged apparently in a fair way to cure.

This case is less clear than the preceding, but the degree of enlargement of the spleen will not account for the œdema, &c., and the general severe symptoms, and it is probably one of anchylostomiasis also. It seems probable that the violent and repeated attacks of diarrhœa may have dislodged the worms. I am strongly inclined to believe that this sometimes takes place, though I have not as yet been able to find worms in such a case. My opportunities, however, have been small, as I have nearly always given thymol after but few days' observation.

Case No. 24, Dachi, a girl aged 12, daughter of a culti-vator; was driven away from the village, because she had *kála-azár*. Her parents had both died of it. She was found on the road, and brought into Tura in a state of semi-starvation.

Present condition.—Body thin, feet and face œdematous, anæmic, bowels loose. Her spleen is very large.

This patient rallied at first, but suffered almost conti-nuously from diarrhœa and severe abdominal pain. Her tem-perature, though not of a malarial type, remained at one time continuously between 101° and 102°. It is only recorded, how-ever, for a few days, and the pyrexia was probably connected with a severe attack of enteritis. She lingered on for three months, and finally died from a fresh bowel-attack.

Post-mortem examination.—Body much emaciated, section of tissues anæmic.

A little serum in pericardium and pleura, but thoracic organs otherwise healthy.

Liver large, somewhat congested.

Spleen very much enlarged.

Duodenum somewhat thickened and large. Jejunum has a darkened appearance caused by a remarkably fine dotted pigmented deposit not ulcerated. There were arborescent patches of congestion in the ilum, but Peyer's patches were healthy, and so were all other remaining organs.

There can be of course no doubt as to the gravity of the malarial complication in the case of this poor waif, but the *post-mortem* appearances leave no doubt as to the case being one of anchylostomiasis, and, as she died from the irritation to the small intestine caused by the parasites, it was evidently this, and not the malarial complication, that caused the fatal issue.

Case No. 25, a Gáro woman, aged 28, wife of a cultiva-tor. Lives at Marapura, where three people had already been attacked by *kála-azár*, one of whom has since died. She was taken ill about a year ago, and since then, her husband has become ill, and her child has died. The illness commenced with fever, followed by enlargement of spleen. Has had diarrhœa for two or three months, severe pain in the head, lost flesh, and has become unable to work.

Present condition.—Is still able to stand and walk, but is very weak and emaciated. Her appetite is now fair. She is anæmic, her feet are œdamentous, and her spleen is enlarged to six fingers breadth below the ribs. Her skin does not appear to have grown darker.

This patient seems to have improved somewhat; she does not appear to have had any fever whatever during her stay in hospital, which extended to two months and a half, but the termination of the case is not recorded, as she appears to have been still under treatment when the collection of cases was sent in. This, too, is in all probability a case of anchylostomiasis, at any rate the amount of splenic enlargement is quite inadequate to account for the anæmia and anasarca.

The remaining 25 cases of Mr. McNaught's series include several which there can be no doubt were instances of anchylostomiasis. They also include one case of acute pneumonia, two of acute dysentery, several cases of ordinary or remittent fever, which of course yielded to treatment, and one or two cases in which the evidence points to malarial cachexia, besides some cases in which apparently anchylostomiasis and malarial cachexia coexisted in various proportions. In some cases from the police and jail hospital there is nothing to show that the patients believed themselves to have *kála-azár*, and they seem to have been classed as such, simply because they had fever.

It is quite evident that Mr. McNaught was on the verge of discovering the true nature of the cases with which he had to deal. The careful examination he made of Peyer's patches in each case shows that he suspected some connection with typhoid fever, but in place of the lesions characteristic of that disease, he found changes in the upper part of the gastro-intestinal canal which he saw were quite peculiar, but which, owing to the non-discovery of the worms that cause them, remained inexplicable. The preliminary washing necessary for a careful examination of the mucous surface necessarily washed away the worms, and left him only the changes produced by them to observe. All his dissections were made several hours after death, or, in other words, at a period when the worms would have become detached from the mucous membrane, and so enveloped in the glairy yellow mucous of the duodenum as to be quite invisible except to closest scrutiny, directed especially to their discovery. Had he chanced to have made one of his examinations immediately after death, while the parasite is still fixed to the mucous membrane, he could scarcely have failed to have found them and to recognize the causal connection between them, and the peculiar, and otherwise inexplicable lesions he had found in those portions of the bowel occupied by them.

Case No. 26 (Dr. Nandi's series).—Manara, an Assamese Muhammadan, aged 25. For the last five years has lived in Gauháti as a domestic servant in a native family. Before this he served as water-carrier, on a tea-garden for about two years.

About six months ago suffered from fever, and since then he has got steadily worse, suffering mainly from dyspeptic symptoms.

Present condition.—Is markedly anæmic and emaciated, with a much enlarged spleen and slight anasarca of the feet. Can walk a little. Urine s. p. 1,010, acid, no sugar or albumen. As he was suffering from diarrhœa, no purgative was given, but he was at once placed upon slop diet and thymol administered, four doses of 30 grains being given at 6 a.m., 8 a.m., 10 a.m., and noon. He passed one stool at 9·30 and another at noon, and both were examined for anchylostoma, as well as one passed at 4 p.m., the previous day, before the administration of thymol, but no worms could be detected.

This and the following four cases are interesting, because they are not originally contributed as examples of *kála-azár*, but in compliance with a circular of the Sanitary Commissioner, requesting an investigation as to whether or not anchylostomiasis existed among Native Assamese as well as among imported coolies. That the case was one of anchylostomiasis there cannot be the slightest doubt, and the way in which the discovery of the worms was missed is a good illustration of the way in which misleading evidence has accumulated. It will be observed that the last stool examined was one passed about the same time as the administration of the last dose of thymol, and only six hours after the first. Food and badly soluble drugs such as thymol introduced into the stomach remain there, as a rule, about four hours, so that only two hours are left for the intestinal contents to pass from the duodenum to defæcation, a time entirely inadequate, even supposing the worms to be destroyed by the first contact of the drug.

I have seldom, if ever, found any worms until some eight hours after the last dose, and it would have been truly surprising if they had been found in this case. The next case, in which Dr. Nandi succeeded in discovering the worms, illustrates this point.

Case No. 27.—Domoho, an Assamese Hindu convict, age about 32, has been in jail about a year, previously to which, with the exception of a month, during which he served in Gauháti as a servant, he had resided at Nomati in the Nalbari thána. About four months ago, had fever and cough, and has not been well since, suffering much from palpitation of the heart, dyspepsia, occasional diarrhœa, alternating with costiveness, with rumblings, and occasional abdominal pain.

Present condition.—Marked anæmia and weakness, with some emaciation. Feet just pit on pressure. After the administration of a preliminary saline purgative, the patient was given three 30-grain doses of thymol at 7 a.m., 9 a.m., and 11 a.m. followed by a second saline purge.

At 5-30 p.m. (*i.e.*, 6½ hours after the last dose), he passed a stool containing 25 anchylostomes.

At 6 a.m. a stool containing only two ascarides.

„ 8 p.m. ⎫ passed stools, which were collected in a single
„ 9 „ ⎬ vessel, and found to contain a number of
„ 10 „ ⎭ anchylostoma and one ascaris.

It will be observed that it is not till nine hours after the last dose that the bulk of the worms appear. It is improbable that he contracted the disease in jail, so that it was probably contracted in his native village, a fact which would seem to point to a somewhat prolonged period of incubation, during which the worms were slowly growing to maturity. My experiments on monkeys, though not conclusive, would seem to point to a similar conclusion.

Case No. 28.—Kikhor, aged 20, a resident of Gauháti, but in the habit of making trips to upper Assam in pursuance of his calling, as a boatman. For the last six months has been suffering from abdominal pain, alternating diarrhœa and constipation, with palpitation of the heart. Is anæmic and weak, feet œdematous, face to a less extent so. Has enlarged spleen, and the hepatic region is rather full, with pain in left hypochondriac and umbilical regions, and winces on pressure over hepatic region. Urine high coloured, no albumen.

Treatment—Calomel, grains ii., every four hours, for four times, followed next morning by thymol, grains xxx, at 6 a.m., 8 a.m., and 11 a.m. He passed one stool at noon, and a second at 1 p.m., which were examined, but no anchylostoma were found.

The patient died four days after, and a *post-mortem* examination was made. The description of the changes recorded

F

reads much like some of Mr. McNaught's cases, with the addition that nearly the whole length of the intestinal mucous membrance was closely scrutinized, but no anchylostoma found.

Here, again, the last examination of the stools was made only an hour after the last dose of thymol, and however many worms may have been ejected by it, they would not be discoverable in the stools until many hours after. After such a thorough dosing with thymol and purgatives, the close search for anchylostomes was hardly likely to be rewarded, as all would be, almost infallibly, cleared out of the intestine.

Case No. 29.—Jugil Kachari, aged 13. Has served for the last three years on a tea-garden, whence he was admitted to the charitable dispensary, Gauháti. About a year ago suffered from fever, and two months after this his feet commenced to swell, and he gradually became very weak.

Present state. Is anæmic, suffers from palpitation of the heart, pulse 100. Spleen enlarged, feet œdematous, face slightly swollen. Appetite bad. Suffers from alterating diarrhœa and constipation. Liver slightly enlarged. Thymol was given in 15 grain doses (a quantity I have personally found inadequate), and the last stool examined was passed five hours after the last dose.

Under the circumstances, it is not surprising that no worms were found. Here is a regular case of tea-garden *beri-beri,* yet observe the identity of the symptoms with those recorded in *kála-azár,* even to the fever and hepatic and splenic enlargement, and notice the double fallacy involved in the assumption that no anchylostomes were present.

Case No. 30.—Shaik Hyat, an Assamese Muhammadan, aged about 40. Has served off and on in tea-gardens, was in hospital five or six months ago with syphilis. Has lately become weak, and has, for a long time, suffered from diarrhœa and dyspepsia. A little while ago had an attack of melœna, lasting two or three days, during which he passed much blood, quite unaccompanied by pain or tenesmus.

Present condition—Emaciated, weak, and anæmic. Feet, legs, and scrotum œdematous; a little ascites. Breathing laboured, appetite bad. No enlargement of liver or spleen. He was given thymol at 6 a.m., 8 a.m., and 10 a.m., and stools passed at 9 a.m. and 11 a.m. were examined for worms, but naturally enough none were found. The case, of course, is a typical one of anchylostomiasis. What other disease indeed will account for the painless melœna?

To this collection I add two of my own cases observed in the charitable dispensary, Gauháti, which illustrate certain points which are not noticed in any of the preceding cases.

Case No. 31.—Api, aged 4, an orphan, can give no account of herself, but is a Kachári, a class whose diet is of a mixed character, and in fact will eat almost anything.

The child is extremely weak; cannot stand without assistance; there is general œdema and ascites; and the face is very puffy, but there is no special prominence of œdema along crest of tibia. There is no sign of paresis or of diminished sensation, and the special senses appear normal. Complexion moderately dark, skin dry and hard. She is anæmic to the last degree, tongue clean, but deadly pale, pulse 102, very weak. Bowels loose, but motion not dysenteric. The lower part of the chest contains fluid, but the lungs and heart are otherwise normal. The liver dulness extends barely to the costal margin: above, its limits are masked by the presence of fluid in the chest. The spleen appears much enlarged, but, owing to the presence of ascites, and much flatus, it is difficult to define its exact limits. The blood was very watery, and poor, in red corpuscles, which did not exhibit the normal uniformity of size. No micro-organisms could be detected in it, and attempted cultivations gave negative results.

A few days after she came under observation, the *post mortem* in which anchylostomes was first found, occurred, and, on examining her dejecta, large numbers of ova were discovered. She was given thymol, and a large number of anchylostomes and a few lumbrici were expelled. No good, however, resulted, and I am even inclined to think that the fatal termination may have been accelerated. She gradually sank, and died about three weeks after she came under my observation. There were a few irregular elevations of temperature, but otherwise the temperature was constantly subnormal.

Post-mortem examination, made three hours after death. Total weight, 30 pounds. Body emaciated, but extremely œdematous, the distended cellular tissue being an inch deep over the chest, and, like all the other tissues, excessively anæmic. Abdomen, pluræ and pericardium full of fluid. Brain 2 pounds; some fluid in ventricles, otherwise healthy. Heart 2¼ ounces; muscle rather œdematous, right side contains a large *ante-mortem* clot, a tail of it extending into the pulmonary artery. Valves healthy, with the exception of three small hard granulations (old) on posterior flap of mitral.

F 2

Lungs—Right 4 ounce, left 3½ ounce. Both very œdematous; a few scattered miliary tubercles on surface of right; none on left.

Liver, 1 pound 5 ounces, of normal appearance, gall bladder much dilated, containing, greenish, glairy bile.

Spleen, 9 ounces, dense, and harder than normal; no adhesions on the surface.

Kidneys, weigh together 4 ounces. Capsule peels easily in both, right with convex border congested, interior pale. Left pale throughout, cortex somewhat atrophied.

Stomach shows scattered petechiœ. Greater curvature of a dull bluish red colour.

Intestine, duodenum and upper part of jejunum of a peculiar slaty hue, closely spotted with marks of old and recent bites. In duodenum and upper part jejunum ten anchylostomes were found (eight females and two males) adherent to the mucous membrane, most being found about three feet down jejunam. Besides these, a number of shred-like bodies containing ova (probably dead, half-digested parasites) were found in the intestine. The ileum contained a number of *lumbrici*, and *trichocephali* were found in the cœcum, some lying coiled up within ulcerated surfaces. Near the anus a number of *oxyurides*.

Case No. 32.—Jagat Das, a Hindustani, who had lived for some years in the neighbourhood of Gauháti. Age about 30. Much emaciated. Height 5 feet 6 inches, weight 90lbs. Can scarcely stand, and is too apathetic to give any clear account of his past history.

Face puffy, legs œdematous. No modification of sensory or motor power, apart from extreme weakness and hebetude. Is frightfully anæmic, and the tongue is very pale, but furred. Skin dry and cold.

Area of cardiac dulness somewhat increased, but difficult to define, on account of being merged below in the dulness extending over the lower pulmonary area. A weak systolic murmur, marked ascites, but no other abnormality of abdominal organs. He is constipated, but complains of no pain. Some kaladana was given to relieve the bowels, and the dejecta were found to contain large numbers of ova, but the food was passed mainly undigested.

This man lived only three days after admission. Thymol could only have precipitated matters, and so was not given, The temperature was persistently sub-normal, and, two hours

before death, was only 90° in the rectum, the lowest temperature I have ever seen recorded in a living human being.
The blood was repeatedly examined at different periods of
the day, but no micro-organisms could be discovered.

Post-mortem examination four hours after death.—Emaciated and anæmic, but very œdematous.

Chest.—A small quantity of fluid in each pleura, the
pericardium contained much fluid.

Lungs.—Right 1¼lb., left 1½lb. Both extremely œdematous throughout, congested below, extremely pale above. The
larger branches pulmonary arteries contained *ante-mortem* clots.

Heart.—Right side filled with an enormous *ante-mortem*
clot, which extended into the auricular appendix, and into
the pulmonary artery and its branches. Muscular substance
thin and pale. Mitral valve and adjacent endocardium somewhat thickened. Left ventricle empty.

Abdomen.—Contained a quantity of ascitic fluid. Liver
1lb. 15½oz., pale, but otherwise normal, gall-bladder moderately
full.

Spleen 5oz., tissue of normal appearance, very firm.
Kidneys each 4oz. Pale, but otherwise healthy.

Intestinal canal.—Œsophagus pale, but normal. Stomach
dilated, pale, with a few reddened patches and petechiæ, the
walls of the organ are so thin as to be nearly transparent. The
bowel was loaded with alvine secretion throughout the upper
part, containing mucous streaked with blood. The ascending
colon containing soft fæcal matter, while the lower bowel is
filled with a hard mass. No ova could be found in the mucus
from the upper part of the bowel; they were extremely
numerous in the soft-fæcal matter in the ascending colon,
while they were comparatively rare in the hardened mass.
The upper end of the small intestine must have contained
about a thousand anchylostomes. The first specimen was
found adherent immediately below the pylorus, and they
were adherent in great numbers to the whole of the duodenum,
and upper third of jejunum. The greatest numbers to be
found in the commencement of the jejunum, where 65 were
counted adherent to a piece of bowel eight inches in length;
on this piece only 10 more recent bites than adherent worms
could be counted. About 3½ feet along the jejunum, only
30 could be counted in a similar length, while at 5 feet
only unadherent specimens could be found, and a few such
could be found throughout the greater part of the length
of the small intestine. No other entozoa were present. The

aorta and larger veins, as well as the thoracic duct, were examined, but presented nothing abnormal.

At first sight, it might appear, from a perusal of this collection of cases, that we were as far as ever from being able to define what *kála-azár* is. ·It will be observed that it includes cases of the most diverse character. For the most part, however, all are chronic cases, whose common leading characteristic is cachexia. Even this characteristic, however, is not universal, for, as we have seen, cases of acute maladies, such as dysentery and pneumonia, are included.

Putting these aside, however, the common symptom that characterizes the greater number of the cases is anæmia, and the prominence and importance of this symptom has been noted by all observers, in illustration of which we may now with advantage add some of the remarks as to the general symptoms of the disease that have been made, from time to time, by medical officers who have given special attention to the subject.

Deputy-Surgeon-General Clarke (Annual Sanitary Report, 1882), epitomizing the accounts he had received from various sources, says " a disease commencing with pyrexial attacks, ushered in with rigors, and followed by sweating frequently recurring, accompanied by pains in the joints and continued headache, resulting in a *condition of general anæmia* with splenic and hepatic enlargement. This is followed by extensive anasarca of hæmic or renal characters affecting the face, eyelids, abdomen, and feet; occasionally by melanasma, epistaxis, diarrhœa, melœna, aphonia, and symptoms of laryngeal and bronchial catarrh. The disease generally terminates by death from general asthenia. * * * Nearly all the preceding symptoms seem to depend on one other symptom not yet mentioned. That symptom is an intense anæmia, which shows itself at an early period, and continues to increase in intensity as the disease progresses."

Now, I have strong reason for believing that anchylostomiasis does often commence in a pyrexial attack, but putting aside the pyrexial onset, could one possibly have a better account of the symptoms of anchylostomiasis than is here given?

Dr. Borah (Appendix to Annual Report of 1888) says —"The train of symptoms appear in the following order :— Anæmia, enlargement of the liver, loss of appetite, emaciation, œdema of the feet, diarrhœa, ascites, and, in a very few cases, cancrum oris (due to poverty of the blood), also enlarged spleen and phthisis pulmonalis."

Here again, the description will apply well to cases of anchylostomiasis and to no other disease. The order of appearance of the dropsical symptoms, too, is given quite definitely and accurately, and is, as I have already observed, the reverse of what is found in malarial cachexia. The very small prominence given to enlargement of the spleen in this description is another indication of the carefulness of the observations that underlie it, especially when we remember that the writer was fully convinced of the malarial origin of the disease, and, but for this predisposition, must have recognised that, though anæmia may result from chronic malaria, it is a sequel, and not a premonitory symptom, of the disease, as here described. It is evident, indeed, that Dr. Borah strongly suspected the true cause of the malady, for he states that he used thymol in some cases, but failed to discover any worms. That he did not find them is probably due to the fact that ·it is quite useless attempting to find them by a mere inspection of the dejecta, unless they be present in quite exceptional numbers, though it is equally possible that the cases on which he happened to try the drug, may have been instances of ordinary malarial cachexia, or other wasting disease.

Dr. P. M. Gupta, speaking of the disease, as seen in the Gáro Hills, in the same appendix, describes the symptoms as follows :—" Obstinate diarrhœa and dysentery, inflammation, affections of the respiratory organs, enlarged spleen and liver, epistaxis, with a general tendency to hœmorrhage, dreadful anæmia, dropsy, and general anasarca are the most prominent complications." Here there is no attempt to describe the order of appearance of the symptoms, but the general description answers most thoroughly to the complications that appear in the anæmia of anchylostomiasis.

To these pictures of the disease I can add but little, and, as I shall have to refer more fully to this point when writing of anchylostomiasis as such, I will not here insert observations, which would necessarily have to be repeated. It is obvious that acute cases of ordinary tropical maladies, differing, as they do, in no way from cases met with all over India, can have nothing whatever to do with the enhanced mortality, and the same remark applies to chronic cases, such as the instances of caries of the spine, of phthisis, and of lardaceous disease which were brought me as *kála-azár.*

Really, the only common character of the cases is that they believed themselves to be suffering from *kála-azár,* and, in point of fact, it will be found in visiting a *kála-azár*-stricken village that any and every case of illness, alike chronic and acute, will be brought out and exhibited to one under that name. Nor is this unnatural. Even the most highly-civilised natives of Europe can hardly be trusted to diagnose their own diseases; and I would submit that if, in all cases, as has been done in this instance, we left the diagnosis to the patient, there are few diseases in which we should not be in equal confusion, and that, for example, in India at least we should draw no distinction between small-pox, measles, and varicella; for it is notorious that, in most parts of India, natives return all three as small-pox: yet either of the three might be the cause of a widely-spread epidemic, and to argue that an epidemic of measles must necessarily be one of small-pox, because the latter disease had always been common in the affected district, and the general populace saw no difference between them, would be obviously fallacious, and none the less so, because it would be necessarily easy to find plenty of cases of true small-pox to support one's contention.

To account, however, for a malady which causes a mortality so rapidly increasing, so severe, and so different from our previous experience of the diseases of the country, we must seek some newly-introduced cause: some disease which has not been always endemic, and not a malady like malaria, which has always been rife in Assam, and which depends on conditions which have remained entirely unchanged. We are, it is true, still much in the dark as to the exact causation of malaria, but, if this enormous increase of sickness be due to it, we ought to be well on our way to a certainty, as to some at least of its effective causes, for these causes should be in prominent action in the affected villages to an extent seldom before realised. Nothing of the sort, however, can be found. I am aware that the disturbance of the drainage lines of the country by newly-constructed embanked roads, and also the extension of forest reserves, have each been suggested as the possible cause of the assumed intensification of malaria; but neither suggestion will bear even the most casual examination.

To begin with, the distribution of the malady bears no relation whatever either to roads or reserves. Further, if an embanked road, by stopping the outflow of water, render the ground above it damper, it is perfectly clear that it must necessarily render that below it drier. Water cannot flow in more than one direction. If such a road be placed in the line of the gradient of the land, it can exert no influence in either direction. If, however, it be placed more or less parallel to the lines of contour, it must render the contours above it damper and those below it drier. Most of the great roads do, in point of fact, run parallel with the contouring of the country, or, in other words, between the hills and the river, and must, in so far, tend to render the land between the hills and the road somewhat damper than before their construction. I doubt, however, if they are adequate to produce any very appreciable effect, for they seem very fairly provided with openings, and *bhils*, and other collections of standing water seem just as common on the one side as the other; moreover, if they really did interfere much with drainage, it is perfectly certain that they would not stand long.

However, assuming them even to have such an effect, and so to enhance the malariousness of the country, on one side of the road, it is perfectly clear that the general health ought to be proportionally bettered on the other side, but no one has even suggested that this has taken place.

In a country like Assam, where so large a proportion of the total area is virgin forest, I have never been able to understand how any one could soberly suggest that the petty amount of afforestation effected by man can have any influence on disease. It seems like accounting for the griminess of Newcastle by the importation of a few baskets of charcoal.

In the absence, then, of any intelligible cause for intensification of the well-known malariousness of the affected districts, it seems as great a mistake to attach any importance, in the production of the increased mortality, to such cases of malarial cachexia as one meets with, as to the cases of dysentery, pneumonia, and spinal caries.

Personally, I have not found cases of uncomplicated malarial cachexia at all commoner than I have in certain parts, for example, of the Central Provinces and Punjab. What I have, however, found is that, wherever a village is seriously affected with *kála-azár*, the larger proportion of serious illness will be found to consist of cases of anchylostomiasis. That a very large proportion of the cases will be found to be complicated with enlargement of the spleen is quite beside the question, for, as we have seen, this condition nearly as often complicates apparent health. Malaria doubtless accounts now for as much sickness and death as it always has, and that, we well know, is no small amount; but it is entirely inadequate to serve as the efficient cause of the terrible death-rate of *kála-azár*-stricken villages, and the true cause of this will be found to be neither more nor less than anchylostomiasis. Of course, it is perfectly intelligible that a man brought low by anchylostomiasis should more readily fall a victim to malarial influences. In the closely-analogous case of famine, we know this to be the case, but it is to shortness of food, and not to malaria, that we ascribe the high death-rate of periods of famine, in spite of the fact that many poor, half-starved wretches might have survived but for attacks of fever and enlargement of the spleen. Conversely, a man, already the subject of malarial cachexia, will die much more rapidly from anchylostomiasis than an originally healthy subject.

As will be seen, I am far from asserting that any and every case that will be produced as *kála-azár* is necessarily anchylostomiasis, or that cases of malarial cachexia are otherwise than very common, for such cases are very common now, always have been, and it will be long ere they cease to be so. All I wish to convey is that the *increased mortality* is due to the anchylostomiasis, and to no·other cause, and hence the answer to the question propounded at the commencement of the section must be that if we take *kála-azár* to be anything brought as such, *kála-azár* may be anything, but that if we confine ourself to the cause of the present pestilence, the reply is that it is anchylostomiasis.

If, as it is to be sincerely hoped will not be the case, the term *kála-azár* be retained at all, it should, I think, be confined to the cause of increased mortality, which has given rise to the term. The preferable course, however, will be to classify our cases under the headings of ordinary medical nomenclature, and to avoid entirely the use of such misleading terms as *kála-azár* and *beri-beri*.

VI.—THE LIFE HISTORY OF THE PARASITE.

The immense numbers of ova which are to be found in the dejecta of patients suffering from anchylostomiasis has already been adverted to. We must now attempt to follow what happens to these after they have been deposited on the ground.

If we watch a deposit, we shall find that, in warm, damp weather, by the next day there will be little or no offence, and little to be seen of the fæcal mass. In its place we shall find the ground turned up into minute heaps of granules. This has been mainly effected by the agency of several species of beetles which feed on the excrement, and carry much of it down into their burrows.

In spite, however, of the destruction of the greater part of the mass by these little animals, if we examine the soil, the next day, *i.e.*, two days after deposition, we shall find it absolutely swarming with minute nematode embryos. In three or four days, the place where the fæces were deposited will be difficult to make out, and by the end of ten days no trace of its presence will remain to the naked eye, but the microscope will reveal the presence of the embryos, in even increased numbers. During the cold season, when many of the species, which thus act as scavengers, are sluggish, or perhaps hybernating, the deposit will remain longer on the ground, and its traces will be discernible for a long time, but the ultimate result, though deferred, will be the same.

As will afterwards appear, when we come to discuss the methods of infection, this rapid disappearance of the offensive matter, and the persistence of the embryos on the site it occupied, has an important bearing on the transmission of the disease. In the case of the small pits, which have been already mentioned, where there is a considerable accumulation of fæcal matter in one place, the efforts of the scavenger beetles are insufficient, indeed they will not be found except at the edges of such accumulations, and hence in this instance the greater part of the fæcal matter remains at the disposal of the nematodes, which are thus enabled to flourish even more vigorously than in the case of isolated deposits.

For purposes of investigation, however, it is better to institute a regular cultivation, but nothing but fæcal matter will serve for the proper nourishment and development of the worms, so that the course of events as seen in nature must be followed as closely as possible.

I have transferred embryos to the ordinary nutritive media, used in bacteriological investigation, to meat juice and other fluids that, it appeared, might offer a suitable and inoffensive cultivating material, but, though the embryos often contrived to live in these for several days, all growth and development was at once stopped. From the circumstance that the intestinal contents of the embryos change to a purple hue under the action of Pettenkofer's test, it may be considered probable that the worms find their nourishment mainly in the biliary matter found in the fæces, and, as far as my experiments go, they show that it is quite useless attempting to observe the development of the worms except they be provided with fæcal nutriment.

However, by utilizing the deodorizing powers of earth, we may obtain cultivations of the worms, which if not absolutely free from offence can yet be dealt with without any great amount of personal annoyance. After trying a great many plans, I find the method most favourable to the development of the worms is to place the suspected fæces in a lightly-covered vessel, such as a *qamlah* tied over with muslin, for a day or two, and then to add sufficient water to render the mass quite fluid.

A shallow crystallizing glass is then about half filled with clean white sand, which should not be too fine, and a sufficiency of the diluted fæcal matter is poured over this, without touching the sides, so as to thoroughly wet the sand without causing obvious pools on the surface; and, finally, the whole is covered with a piece of clean window glass. From time to time (every day or two) the glass cover must be raised so as to renew the air within the chamber, as I find that the worms can only develop in the presence of a free supply of oxygen. Freshly-collected fæces may be employed in the same way, but I am inclined to think, as a result of many experiments, that the development is rendered more sure by the ova being allowed to remain from 24 to 48 hours in their natural habitat.

A certain amount of dampness is important in ensuring the best result, and the degree of this may be best expressed by stating that the grains of wet sand should still hold a certain amount of air entangled amongst them. No doubt, this is connected with the necessity the worms, in common with other animals, have of obtaining a free supply of oxygen. The length of time taken in hatching out varies greatly with the temperature of the air. At what degree development altogether stops I am unable to state, as, owing to the impossibility of obtaining ice, I was unable to institute any experiments on the point. As far, however, as Assam is concerned this is unimportant, as the temperature here, even in the cold weather, never falls sufficiently low to do more than retard the hatching process.

In December, at Gauháti, the temperature in my laboratory ranging a little, above 60° F. during the day, and falling to about 54° at night, the embryos do not commence to hatch out until the fifth day, and then only a very small number, the greater proportion not appearing until the sixth day. On the other hand, in an observation made at Dhubri late in March, the temperature being about 84° in the room where I was working, the embryos hatched out the day after the deposition of the excrement; while in Shillong, at an intermediate temperature of about 70°, they usually did not appear till the third or fourth day. I made some attempts at ascertaining the effects of a higher temperature, but my constant temperature apparatus proved a failure, and the experiments that were attempted in it were necessarily inconclusive. They showed, however, that the embryos can live for a considerable time at a temperature of 120° F.

The ova, as met with in the fæces, have a very characteristic appearance, and once seen cannot well be mistaken for anything else. As the average of a large number of micrometric observations, I find they measure $\frac{1}{420}$ inch in length by $\frac{1}{630}$ inch in breadth. As far as can be made out from the very rough figure in Dr. Kinsey's pamphlet, this coincides pretty closely with Perroncito's measurements, but his figure, to judge from the varying size of the ova represented, can only be a free-hand production, and therefore cannot be taken as exact. I can find no statement as to the measurements of the ova in the papers of either Dr. Adolph Lutz or Dr. Schulthess.

In the last edition of Dr. Bristowe's "Theory and Practice of Medicine," the measurements are stated as $\frac{1}{540}$ inch by $\frac{1}{1000}$ inch. The statement is of course, not original, as the author could have had no opportunity of personal observation, and the authority for it is not given, but if it be based on correct observations, it must refer to a distinct species from that observed in Assam, as the measurements of the ova of any one species vary within such narrow limits only, that it would be impossible to find any two ova differing by as much as $\frac{1}{420}$ inch and $\frac{1}{540}$ inch, so that either the ova referred to belong to different species, or some mistake has been made by the authority from which Dr. Bristowe took his description.

The exact measurement, however, is of great importance in the diagnoses of cases by microscopical examination of the fœces, especially when it is attempted for the first time, before one has become familiar with the general appearance of the ova ; and hence this statement in Bristowe gave me a great deal of trouble in the early stage of the investigation, and, mistrusting the exactness of my stage micrometer, I sent for a new one, but both instruments give the same readings, so that the above measurement may be accepted as exact.

In fresh fœces the ova are usually 2-4 segmented, but I have often met with unsegmented ova. The yelk is of a greyish colour, and is separated from the shell by a comparatively wide zone of perfectly clear, transparent fluid. As this fluid and the egg-shell are alike quite white, it is by this character alone perfectly distinguished from the ova of *Ascaris lumbricoides*, which are yellow, and from those of *Trichocephalus dispar*, which are deep red brown. This clear zone is usually wider at the poles than elsewhere, and is not found to the same extent in any other human parasite. Another peculiarity is that the egg has a tendency to a cylindrical form ; in other words, its outline is not truly elliptical, but has its longer sides somewhat flattened.

The only human parasite with whose eggs it can possibly be confounded is the *Oxyuris vermicularis*, but this latter is somewhat smaller, being only $\frac{1}{400}$ inch in length, its outline is unsymmetrical, being flattened on one side only, and, lastly, it always contains a well-advanced embryo. Now the most advanced *Dochmius* ova which can be found in fresh fœces are in the morula stage of development, so that the mistake could only arise in examining fœces that have been passed some time, though, as we have seen, in hot weather 18 or 20 hours would be sufficient.

, After reaching the morula stage, the first indication of the embryo appears as a slight groove, rather to one side of the pole of the morula. This groove deepens until the embryo can clearly be made out as a tadpole-shaped body bent on itself. The embryo now rapidly increases in length, at the expense of the thickness of the larger end, until a distinctly worm-like body results. Before this process is completed even, slow movements begin to be observable in the mass, until at length the embryonic worm can be seen in constant motion coiling itself about in the egg with such rapidity that it is very difficult to obtain a correct drawing. The movements become more and more violent, the creature appearing to struggle to straighten itself out, till at last the shell ruptures, as a rule, a little to one side of one of its poles. The little worm soon wriggles itself free from the empty shell, and at once begins to move about actively in the fluid in which it finds itself.

Immediately after emerging from the egg, the embryo measures 0·085″ in length by 0·005″ in diameter. It tapers a little towards the head, which is rather blunt, and terminates in a finely-pointed tail. The whole length is marked with fine transverse striations as close as those of muscle fibre. Under a high power, the intestinal canal can be clearly made out, but is not as prominent an object as it becomes a few hours later, when its endothelial cells have become plumped out with granular matter.

The commencement of the intestine proper is about one-fifth of the entire body length from the oval end of the body. Immediately in front of this is a small chitinous bulb, which is armed with certain chitinous ridges, so disposed as to appear like the mouth of a leech, when seen in optical section. When first escaped from the egg, this bulb is in a continuous state of pulsation, its strokes being about 60 per minute, but not absolutely regular, sometime pausing for a while. It is no doubt by this means that the large supply of nourishment required for its growth is drawn into the body. After an hour or so, these pulsations become slower, and finally cease. In front of the bulb, the œsophagus can be traced as a faint outline, broadening somewhat at about half its length, and extending to the mouth. The little animal grows at first with extreme rapidity. Within the next 48 hours, at an average temperature of 70°, they have increased to three times their original length. During this time, they have undergone several ecdyses, but the exact number of changes is difficult to state, as it is impossible to keep any individual worm continuously under observation. Repeated experiments have convinced me that observation of these embryos, under laborating conditions, are quite valueless. The nature of the medium in which they will alone flourish quite precludes microscopical observation, and, in water, development is at once modified and soon arrested. I have made many attempts to induce them to grow in transparent media, but quite fruitlessly. In nutrient beef jelly, such as is used for bacteriological work, whether acid, neutral or alkaline, they soon die, whether it be allowed to solidify or kept fluid by heat or dilution. They live longer, but will not flourish, in urine, and the nearest approach to success was obtained by mixing fæcal matter with water, and filtering through nainsook, whereby a fluid is obtained which, in thin layers, is still clear enough for microscopical observations; but even in this I have quite failed to observe continuously, from birth to the adult rhabditis stage, to be described below, any individual embryo.

It is no doubt owing to too great trust in observation conducted under purely laboratory conditions, that the description of the life history of the worm given in Dr. Adolph Lutz's paper is full of fallacies. As those who have read that otherwise very able paper will remember, a great feature is made of what he terms the encapsuling and calcification of the embryonic parasites. I can emphatically assert that in their life history there is no such phase, under any conditions whatsoever; and that what has been so described is not life, but death history, in other words, they are but the changes that occur in dead worms when left in water. I have kept dead embryos which presented all the changes figured by Lutz under observation for long periods, and I have never witnessed the least sign of returning life; further I have followed the further change of decomposition into mere granular matter. The only way to successfully observe the normal life history of the parasite is to make a large cultivation and remove small specimens at frequent intervals for microscopical examination in water. It is here that the use of coarse sand as a basis proves so useful, as the sand entirely does away with the tenaciousness of the original material, and a specimen removed to a glass tray, and played upon by a jet of water, falls to pieces, and is ready for observation at once. It may be well, before proceeding further, to describe more minutely the process employed in examining the cultivations. The best size of tray is about 3 inches by 2 inches, internal measurement, and about $\frac{1}{3}$ inch deep. Such tray can easily be made by any one as follows :—

Cut a strip of window glass $\frac{1}{3}$ inch wide, and as thick as can be obtained. Cut off it two pieces, each 3 inches, and two pieces 2 inches long. Cut also, from a thinner piece of glass, as free as may be from flaws, a plate, $2\frac{1}{2}$ in. × $3\frac{1}{2}$ in. Now make one of the long strips hot in a spirit lamp, and smear one edge with shellac, or, better, a mixture of equal parts shellac, yellow beeswax, and resin : now heat one of the long sides of the larger plate, and then apply the cemented strip, so that it is fixed along the edge of the plate, standing upright. Next, in the same way, apply the cement to one edge and one end of each of the shorter strips, and fix them in position in the same way one by one. Lastly, cement one edge and the end of one face of the fourth side, and place it in position, and the tray is completed. I describe the method of making these trays thus minutely, as their use is of some importance in the diagnosis of doubtful cases, as, owing to the lively movements of the embryos, it is much more easy to detect them in a cultivation than to find the ova in the fæces. Moreover, a simple pocket lens, of one inch focal length, affords quite sufficient magnification, so that a comparatively large amount of material can be searched over in a few minutes. A good light is necessary, however, so that, unless a regular dissecting stand be available, it will be necessary to improvise one, to do which the tray should be placed over a hole in one side of a cigar-box, the lid being removed, and a shaving-glass, arranged at a suitable angle, placed within the box. The lens may be supported on a pen-holder fixed in a hole in the back of the box and thrust through the small holes which are found in the sides of the horn case of pocket lenses of the usual pattern. Such an apparatus can be constructed by any one, however little of an amateur mechanic he may be, and forms by far the most certain method of diagnosing the presence of the parasite, when present in only small numbers, and by its means I have again and again detected the presence of a few parasites where repeated careful examination of the fæces for ova had failed. For a reason that will presently appear, the cultivation should be allowed to stand for six or eight days before condemning it as sterile.

On account, then, of the impossibility of keeping the embryos under continuous observation, I am unable to afford any exact information as to the number of ecdyses, but where development is rapid, I estimate the changes of skin at not less than twice a day. I have often watched the process, and some very curious appearances occasionally present themselves, as the worm often remains for a considerable time partially entangled in the old skin.

In every instance that I have watched, the old skin is ruptured near the head, a small cap of the anterior part of the chitinous investment being forced off, while the rest of it is cast off entire, the worm drawing itself out like the finger from a glove, the tail being the last part freed.

Even in one and the same cultivation, the rate of growth

G

varies greatly, depending no doubt on variations of the amount
of available food supply in different portions of it. In experiments conducted at Shillong, under conditions as nearly as
possible approaching those found in actual practice, the whole of
the stages from ovum to the adult rhabditis stage, to be presently
described, occupied six days. In this case the fæcal matter
was simply spread over a patch of garden soil enclosed within four planks. In experiments conducted in glass vessels,
with a necessarily somewhat restricted supply of air, the development takes a trifle longer. The most important factor,
however, is the proportion of fæcal matter, or, in other words,
of nutritive material used. In large fæcal masses, in the
temperature of the plains, fully developed rhabdites may be
found as early as the fourth day. On the other hand, immature forms may still be found as late as the tenth or even
twelfth day after hatching out.

The first change that takes place is that the cells lining
the intestinal canal proper swell out and become granular,
bulging out into the lumen of the channel, so that its course,
as seen in optical section, appears zigzag. Certain of the cells
at its commencement, i.e., immediately below the hinder bulb,
take on a special development, and remain larger than the
rest throughout life. They probably function as a liver.
After one or two changes of skin, the anterior dilatation of
the œsophagus becomes much more distinct. Between it and
the hinder bulb, a nervous centre, consisting of four or five
ganglion cells, encircling the œsophagus, makes its appearance, and lastly, about midway along the ventral aspect of the
body, there appears a single hyaline cell, the rudiment of the
future generative organs. In these early stages, the mouth
is a simple, somewhat funnel-shaped opening, without either
lips or armature of any kind. The anus opens on the ventral
side of the body, a little in front of the pointed end of the tail.

Almost up to the assumption of the adult rhabditis stage,
the changes occurring at each change of skin are very gradual,
and consist merely of growth of the parts that have already
appeared, no further change taking place in the development
of the generative organs until this great and final stage.

In these final changes the worms become adult male and
female organisms, and immediately commence to propagate the
species.

For the benefit of such of my readers as may be unacquainted with the details of nematode terminology, it may be
well here to explain what is meant by a rhabditis stage.

Its meaning is that certain parasite nematodes* present the phenomenon of alternations of generations, the ova laid by worms in the parasitic stage, developing, not into animals like themselves, but into a form which no naturalist would place even in the same genus. In fact, if we were ignorant of their origin, these forms would find their place, for purposes of classification, in Dujardin's genus *Rhabditis*. The worms in this stage live free, and are in no sense parasites, but when they reach sexual maturity, they give birth to a generation which again takes up a parasitic mode of life, or, at any rate, is capable of doing so, while, as far as can be seen, the first generation would be incapable of becoming parasitic.

Thus, the worms of such species, in the parasitic stage, are not the children, but the grandchildren of parasites. In the present instance, the free or rhabditis stage fits with some difficulty into Dujardin's definition of the genus *Rhabditis* as its strongyloid origin is indicated, in the male, by a distinct copulatory bursa supported by chitinous rods, much resembling, though not identical with, that of the parasitic stage. As, however, Dujardin's definition includes certain species, the males of which have the "tail provided with membranous wings," its limits are not much forced, and there is no need to institute a special term for the free stage of our present species. As far as can be made out, the commencement of the sexual differentiation of the male rhabdites takes place in the last ecdysis, but one preceding that in which completely adult characters are assumed. At this period, the worm measures about $\frac{1}{38}$ inch in length by about $\frac{1}{1000}$ inch at the thickest part of the body, opposite the still rudimentary sexual gland, which is placed rather nearer the caudal than oral extremity, the middle of the fusiform rudiment being about $\frac{1}{100}$ inch from the caudal extremity. At this stage the generative gland consists of only one or two hyaline, multinucleate cells. The copulatory spiculæ however, are well developed, and do not change to any extent during the remaining ecdyses. There is, however, no sign of the copulatory bursa, which appears after the next change of skin, the tail being very long, and tapering so as to form a sort of flagellum. Even after the next ecdysis, there is but little change in the generative gland, which, at this period, consists of a hyaline, fusiform body, containing a few clear nuclei,

---

* The number of species in which such a life history has been proved is constantly increasing. Only lately the fact has been made out in the case of *Strongylus strigosus* and *S. retortæ formis* by Prof. A. Railliet, Bull. Soc. Zool. France, XIV. (1889) page 875-7.

placed ventrally, and measuring $\frac{1}{400}$ inch long by $\frac{1}{2500}$ inch in diameter. The caudal extremity is rather abruptly conical, and now shows a well-developed copulatory bursa, consisting of a delicate transparent membrane, shaped like a finger-nail, and projecting laterally a good deal more than it does beyond the point of the obtusely conical tail. It is supported on either side by three simple chitinous rods of uniform diameter, which, taking their origin in the cuticle of the tail, pass from it, outwards and backwards, quite to the edge of the membrane. The width of the entire organ barely equals that of the greatest diameter of the body, and it extends up its sides only $\frac{1}{500}$ inch from the extreme point of the tail. Just above this point, on either side of the middle line of the body, may be seen the copulatory spiculæ, consisting of two pointed chitinous rods, of a deep yellow-brown colour, and about $\frac{1}{1000}$ inch long. Their bases are placed at some distance apart, but their points, when retracted, touch, so as to form a V, each being contained within a separate sheath, the two sheaths, however, joining to form a common opening, which receives also the mouth of the duct of the generative gland, and together open into intestine. close to the anus, to form a short cloaca. The form of these spicules differs greatly from the copulatory organ of the entozoic stage of the worm, in which it is, proportionally, much longer, and as it does not greatly exceed it in actual thickness, is a much slenderer structure. In the entozoic stage, too, the two spicula are kept in habitual contact, so that they can only be separated by dissection, appearing to be, in some way, locked together, and they end in a very minute but distinct hook. These spiculæ are protrusible, and may be seen in different specimens in a variety of positions, and when seen in profile, often appear like a couple of hooks projecting from the ventral aspect of the tail. In other respects the worm as yet differs but little from the younger embryos already described, almost the entire body cavity being filled by the intestinal canal, the excretory gland, which forms so prominent an object in the mature stage, being not yet sufficiently developed to be easily made out.

When fully developed, the mature male rhabdites measures $\frac{1}{30}$ inch — $\frac{1}{24}$ inch in length, but little exceeding in this matter the size of the worm during its two preceding changes of skin. It has, however, become markedly thicker, its greatest diameter, which is still placed in the same situation, reaching from $\frac{1}{500}$ inch — $\frac{1}{300}$ inch. The structure of the nutritive organs remains unchanged, but, lying beneath the ventral aspect of the intestinal canal, may be seen a glandular body of considerable relative size, measuring $\frac{1}{170}$ inch in length by $\frac{1}{1000}$ inch in diameter. Behind it ends in a blunt rounded extremity, while in front it tapers off to form a duct. It is lined with an epithelium of very small, nearly spherical, nucleated cells. In one or two instances I have doubtfully succeeded in tracing the duct of this gland to near the excretory pore, and have little doubt that the organ has an excretory function.[*] It certainly forms no part of the generative gland, the anterior end of which it overlies. On casual examination, it appeared possible that it might be a doubled-over end of the generative gland, but more careful focussing showed that this could not be the case. Were it a reflected position of this organ, the blind end would contain the least-developed cells, and the latter would become progressively advanced in development as one proceeded to the assumed point of reflection, whereas the contrary is the case, the blind end being filled with well-developed cells, while near the point where the faintly-seen duct commences, the cells become granular and degenerated. It is present in adult males and females alike, and is especially easily made out where the animal has been killed by the cautious addition of alcohol, just before the protoplasm becomes opaque from the action of the re-agent. The histological investigation of these organisms is excessively difficult, as, owing to the resistent nature of the chitinous cuticle, the internal organs cannot be acted upon by re-agents sufficiently soon to forestall those early putrefactive changes which are fatal to the production of really instructive preparations.

Owing to their small size, it is impossible to slit them, or make openings, as one would do in the case of larger organisms, and the method of sectionizing is equally impracticable, as, by the time they have passed through the various stages of hardening and imbedding, nothing but a chitinous ring remains. Fortunately, however, their great transparency goes far to neutralize these disadvantages.

The generative gland, which has now definitely assumed the characters of a testis, consists of a single tubular body, commencing a little behind the second pharyngeal bulb.

---

[*] While the above was passing through the press the August 1890 number of the " Journal of the Royal Microscopical Society " came to hand. It contains (page 461) a note of a paper by Dr. O. Hamann in the Zool. Anzieg. XIII. (1890) page 210-2, in which this organ is clearly described ; and it is stated that there is no doubt about the existence of an external opening, and that the organ is a lemniscus homologous with those of the Acanthocephala, and therefore, as I surmised, has a probably excretory function.

At its anterior end, it exhibits, in optical section, a double row of clear, finely granular cells, each possessing a single nucleus and nucleolus. As one proceeds backwards, these cells become larger, until, about half way along the length of the body, they attain so large a size that they can only lie in a single row. The most anterior examples of these larger cells contain only a single nucleus, but further back they become multinucleate, and, by careful focussing of the most developed, a distinct brood of daughter cells can be made out within them. It is within these latter that the spermatic elements take their origin, the whole process being one of endogenous cell multiplication. Finally, the large cells burst, and the hindmost part of the generative duct will, in perfectly mature specimens, be found to be filled with the spermatozoa, which are small rod-like bodies thicker at one end than the other, and so tending to the well-known hat-shape so often seen in nematodes. They are capable of ameboid movement, but are destitute of any lash. As the little creature grows older, the proportion of its length filled with the large single cells increases, while those arranged in double row diminish, until, at length, the whole of the anterior portion of the body is filled with the large mother cells only. This is, of course, owing to the progressive development of the cells, and to the circumstance that no new ones are formed to take up the place of those that have attained complete development. When all its spermogenic cells have completed their development and discharged the spermatozoa to which they have given birth, the rhabdites die. The copulatory bursa of the fully-grown rhabdites, though agreeing in general form, differs a good deal in detail from that of the immature form found before the last change of skin. In this stage, it will be remembered, there were, on each side, but three simple supporting ribs. In the adult stage, there are six ribs on either side, of which, counting from the front, the first, fourth, and sixth alone reach out to the edge of the membrane. The second is the shortest, not reaching half its breadth, while the third is a good deal longer. The fifth attains very nearly to the edge of the membrane, and has the additional peculiarity of being bifurcated to about half its length.

I have never succeeded in finding a pair in copula, but the female is probably grasped in much the same way as in the entozoic stage. It is generally supposed that, in this latter stage, the grasping action of the copulatory bursa is assisted by a viscid secretion, and this view receives support from the structure of the caudal end of the rhabditis stage, in which are to be seen a number of cells of unmistakably glandular character, which may possibly be concerned in the secretion of some such fluid. These cells are situated absolutely in the point of the tail, and so are quite beyond the end of alike the intestine and the generative duct.

In the female rhabdites the changes closely follow those that take place in the male, though they are not so externally prominent, on account of the absence of the copulatory bursa. The generative gland is, for a long time, exactly alike in both sexes, but it makes its appearance nearer the oral end of the female than in the male, being placed a trifle nearer to it than to the aboral extremity, but, beyond this, there is nothing to distinguish it from a male. At a stage corresponding to the appearance of the copulatory spiculæ in the male, the female external generative opening makes its appearance. It is a small oval aperture placed on the ventral aspect of the body, rather nearer the head than the tail, and is guarded by a thickened chitinous ring. The number of changes of skin between the commencement of sexual differentiation and maturity appears to be the same in both sexes, but, up to the last change, the appearance of the internal generative cells is closely similar; only, at all stages, the furthest developed sexual products, instead of being found, as in the male, towards the caudal end of the animal, are placed opposite the opening of the duct, near the middle of the body, and grow less and less mature as one proceeds, on the one hand, towards the tail, and on the other towards the mouth.

The single ovarian tube, in fact, extends in both directions from the genital opening, and hence the least developed germs may always be found in two situations near the head, and the tail respectively. Instead, however, of developing into the mother cells of spermatozoa, the germ cells in this case develop directly, without any process of endogenous multiplication, into ova. The ova produced by the rhabditic stage are in all respects exactly similar to those of the entozoic, which have been already described, and those first extruded are laid in much the same stage of development as those met with in the fæces of infected persons, viz., 2-4 segmented. During the process of laying, the female remains almost motionless, and does not, I am inclined to think, take any food after the completion of maturity. As more and more eggs are laid, the more backward germs, from either extremity of the ovarian tube, advance towards the middle of the body, and develop, in their turn, into distinct ova, until at length, owing to the emptiness of the intestinal canal, she becomes from end to end little else than a sausage-shaped bag of ova. The development of these latter, however, proceeds at a somewhat more rapid rate than the hatching process, so that soon the eggs, when hatched, already contain living embryos, instead of a merely segmented yelk. Finally, when the female is becoming nearly emptied of eggs, they hatch out within her body, and appear to feed on her organs, only gaining access to the exterior by the rupture of her body-wall, by decomposition. Thus, this animal presents the peculiar phenomenon of being oviparous in her young maturity, and ovoviviparous in her old age, and, as may be seen, the completion of the process of laying necessarily involves her death also. As they do not appear to require food after attaining maturity, the whole process can be easily watched in a cell containing water, and I have followed it throughout in several instances. The number of eggs produced, is considerable, but is quite insignificant in comparison with the myriads produced by the entozoic stage of the creature. The largest number observed to be laid by any one female after isolation in a watch-glass was 70, but the necessity of being sure of complete maturity prevented one's being able to select them at the absolute commencement of the process, so that, all considered, we shall probably be not far wrong if we double this figure as a fair average of the family produced by a single female. As has already been pointed out, the process of reproduction absolutely necessitates the death of the female, and her whole life does not much exceed seven days.

In examining impregnated females, the copulatory spicula of the male may often be seen sticking in the opening of the generative duct, so frequently so indeed, that this mutilation of the male is probably the rule, and, if this be the case, it lends strong support to the idea that the life of the male is of no longer duration than that of the female, as, after such a mutilation, his functional utility would necessarily cease. The female is considerably larger than the male, sometimes reaching the length of $\frac{1}{15}$ inch and a breadth of $\frac{1}{200}$ inch, so that with strong oblique transmitted light, they can be made out with the naked eye when once their position has been ascertained under a lens. The tail of the female is acuminate, a small sharp process projecting from the otherwise blunt tail. In the remaining points of her anatomy she exactly resembles the male.

In both sexes, the margin of the lipless mouth is armed with a circlet of very minute papillæ, which, being rather sharp-pointed, might well be mistaken for teeth. The nervous system is best studied in large living specimens, which have been rendered motionless by immersion in water containing as much thymol as it will take up. I found this re-agent extremely useful for this purpose, as it stupefies without killing, and it is quite sufficient to quiet the naturally sluggish adult rhabdites. It is not, however, sufficiently powerful to stop the violent motions of younger embryos, for which purpose, about 4 per cent. of absolute alcohol must be added to the mixture.

The structure and position of the central ganglion has already been described, but the distribution of the nerve cords issuing from it remains to be noticed. Several (6 or 8) fibres proceed forwards, along the sides of the anterior pharyngeal bulb, to the neighbourhood of the mouth, and on these fibres are found a number of irregularly disposed ganglion cells, but little inferior in size to those composing the central ganglion. On each side, a single chord proceeds backwards, and these, too, are provided with irregularly disposed ganglia, but they cannot be traced far, as, owing to the crowding of parts below the second pharyngeal bulb, they are soon lost to view, but, in the clearer space near the tail, cells that seem to have a ganglionic character can again be made out, and there can be little doubt that these are connected by commissural strands with the œsophageal nerve ring.

The embryos hatched out from the eggs of the rhabdites are identical in size and appearance with those already described as hatched out from the eggs produced during the entozoic stage of the worm, and it is entirely a matter of chance whether they will develop into free organisms like their parents, or become parasites like their grandparents.

In all probability, the generation immediately produced from the eggs of the entozoic stage would prove harmless, if introduced into the human intestinal canal, as the life cycle would be incomplete, were they to develop into parasitic nematodes like their parents. There can, however, be no doubt of the infectiveness of the progeny of the rhabdites, and it is probably universally by the agency of these that infection is effected. Attempts have already been made, by other investigators, to infect sheep, pigs, and other domestic animals, but without success, so that it was clearly useless to make any experiments with these, and accordingly the small short-tailed monkey (*Macacus rhesus*), very common in Assam, was selected as a subject for experimentation, but, as will be seen, only partial success has been met with. The first experiment con-

sisted in feeding a young specimen of this species with food
with which was mixed mud that teemed with adult rhabdites
and their progeny.

In the course of a few days, the little animal became
obviously ill. It vomited repeatedly, was continually drinking,
and distinctly feverish ; its temperature being raised to 102°F.
It grew worse, and was obviously sinking, and, as I had to
leave Gauháti to go on tour, I killed it by a heavy dose of
chloroform, and examined the viscera. All organs were perfectly healthy, with the exception of the intestinal canal. The
stomach was patchily reddened just as seen in autopsies on
cases of anchylostomiasis. At first sight, the duodenum seemed
destitute of parasites, but, on placing pieces of the mucous
membrane in a tray of water, under the simple microscope,
countless multitudes of very small nemadotes were seen adherent to the mucous membrane. They had attained, in some
cases, nearly half the length of an adult *Dochmius duodenalis*,
but were extremely slender, the generative organs being
still quite indeterminate, and the whole organisation of the
parasite more like an over-grown, immature rhabditis than an
entozoic *dochmius*, only one or two being sufficiently developed
to admit of the determination of sex, and these were females.

For the rest, the remainder of the intestine swarmed with
other parasites. These were twenty specimens of an *Ascaris*,
which I believe, is identical with the human *A. lumbricoides*,
large numbers of an *Oxyuris*, to all appearance, also identical
with that infesting the human subject, a number of *Trichocephalus dispar*, and one or two specimens of a strongyle worm
which has not as yet been identified. Ova corresponding to
each of these four species had been found in the fæces, previously to the infection of the monkey by the *Dochmius* embryos.
The ova of the strongyle would alone have presented any
difficulty, had the experiment been carried further, as its eggs
are so like those of our species that it would have been practically impossible to distinguish them by microscopical examination. Meanwhile, a second monkey of the same species had
been fed in the same way. Its dejecta contained the ova of
the *oxyurid* parasite alone, so that it was better suited for our
purpose than the first. I had to leave Gauháti before the
animal showed any signs of ill-health, and it died a fortnight
after my departure, *i.e.*, less than three weeks after the first
feeding. Unfortunately, Dr. Mullane was also absent from the
station at the time, and the Hospital-Assistant, in whose charge
the animal had been left, neglected to preserve the intestine in

spirit for my examination. Naturally enough, he saw nothing of the worms in the duodenum, as they can only be made out by careful examination under a lens, so that he found only the *Oxyurides*, but there can be no practical doubt that the animal died from the effects of the invasion of the immense numbers of embryos it had been made to swallow. The extent of development was quite as great as could be expected in an experiment that had lasted so short a time, so that I felt little doubt of being able to rear mature parasites whenever time for the experiment should be at my disposal. After arriving at Shillong, three more monkeys, of the same species, were obtained and fed in the same way, except that a much smaller amount of infected earth was administered, in order to avoid the death of the animal before the parasites had had time to reach maturity. No symptoms beyond occasional mild diarrhœa have been observed in these animals, though they have failed to get into as good condition as animals, as fully fed as they have been, might have been expected to do. In all three, the ova of *Trichocephali* were found in the fæces, as well as those of the oxyurid, but none of those of *Ascarides* or of the strongyle could be found in either. The fæces were examined at intervals, but, in neither case, have the ova of *Dochmius duodenalis* made their appearance. Three months after the feeding, one of the monkeys was killed. The animal was quite healthy, except that, just as in the previous case, numbers of nematodes were found adhering to the mucous membrane of the duodenum, though not of course in anything like the numbers that were found in the first experiment. They had increased a good deal in length, but were otherwise but little in advance of those found in the previous experiment. The sex of all specimens could, however, now be made out, and it is a very curious circumstance that over thirty specimens which were examined microscopically were all females, and a careful search for small specimens entirely failed to bring to light a single male. The remaining two monkeys, though four months in all have elapsed, are still without any sign of *Dochmius* ova in their dejecta, so that there does not appear to be any prospect of the worms with which they have been infected reaching maturity.

Only two interpretations can be put on these later experiments, as far so they have gone : either the embryos of *Dochmius duodenalis* take an extremely long time to reach maturity, which is, on the whole, improbable, or else the parasites are unable to find a sufficiently congenial home in *Macacus rhesus* to admit of their attaining maturity. Personally, I incline to the latter explanation, as it is entirely in accordance with the general phenomena of parasitism, that a parasite peculiar to one animal should be incapable of development in another and different host. Such being the case, I fear that nothing can be hoped for from further experiments on animals. The circumstance that three parasites probably identical with species infesting human beings, viz., *Ascarus lumbricoides, Oxyuris vermicularis,* and *Trichocephalus dispar* find a congenial host in *Macacus rhesus* shows that the latter is as favourable a species for our purpose as we can expect to obtain, though it is possible that better success might be met with by employing some of the higher, tailless apes, the difficulty of obtaining monkeys of such species here precludes the attempt being made at present. Had I met with these worms, without being aware of the feeding to which the monkeys had been subjected, I should certainly have referred it to a genus quite distinct from either *Dochmius* or *Rhabdites,* and it might be suggested that the worms found are simian parasites, having no connection whatever with the *Dochmius* embryos introduced into the intestinal canal of the monkeys experimented upon, but the evidence, I think, is fairly conclusive · that this is not the case, In the first place, the specimens found in the first experiment, where only a short time had elapsed between the feeding and the examination, were distinctly less advanced in development than those found in the experiment where a period of three months had been allowed to intervene; in the next, there was an obvious relationship between the number of parasites found and the severity of the feeding; and, lastly, the immaturity of all specimens found is strongly against its being a separate species, as in nearly every case where a parasite passes any portion of its life-cycle in the intestinal canal, it is during this period that it attains sexual maturity, so that it is quite exceptional ever to find large numbers of immature worms free in this situation, and it would be in the last degree improbable that, under normal circumstances, all should prove to be so. The most developed specimens contained, it is true, fairly developed ova, but none of them gave evidence of having been impregnated, and no ova corresponding to them have ever been found in the dejecta of the animals experimented on, so that I think it will hardly admit of doubt that we have to do with a case of arrested development, and that, in the worms so found, we have a phase of the metamorphosis of the rhabditis embryos, so to speak, crystallized into a permanent condition, the abnormal host affording conditions sufficiently favourable to admit of the commencement of the developmental process, but not affording sufficient, or sufficiently suitable, nourishment to allow the process to proceed beyond the stage we have met with.

While, however, not.entirely conclusive, these experiments go far to prove two-points, first, that the normal method of infection is. through the agency of rhabdites progeny, and, secondly, that the worms fasten on to the intestine at once, for, though the worms have as yet no oral armature like that of the adult, yet they were always found actually adhering to the mucous membrane. The mechanism by which this is effected is not quite clear, but it must be remembered that the oral end of these worms is far sharper and finer than the smallest needle ever manufactured, and that the soft structures of the intestinal mucous membrane can offer but little resistance to such an organism, boring its way into it, simply as a sharp body. That this is easily possible is shown by the way in which *Trichocephalus dispar* bores into the mucosa of the cœcum, for this worm, though equally unarmed, and decidedly larger and coarser at the oral extremity, as is well known, burrows for nearly an inch under the mucous membrane of the gut. Probably, like this latter worm, the little nematode, at this stage, feeds upon the serous fluid permeating the tissues, and not upon the blood, as they do at a later stage; at any rate I have found no evidence that blood forms any part of their food. The first experiment is also instructive, in showing that the invasion of a large number of the parasites may give rise to a distinct febrile attack, and as, in actual practice, infection probably takes place, in most cases, not once only, but repeatedly, it shows how feverish attacks may form a prominent feature during the earlier stage of anchylostomiasis and therefore of *kála-azár*. As to the further development of the parasite, it is open to doubt whether it undergoes all further changes within the intestine, or effects a temporary encystment in the wall of the bowel. In Dr. Schulthes's paper, translated by Dr. J. D. Macdonald in Dr. Kinsey's pamphlet, there is a note to the effect that Griesinzer and Grassi have found specimens encysted in the sub-mucosa. There would be nothing astonishing in such a habit, as the very closely allied *Sclrostonum tetracanthum*. Dies., which infests the large intestine of equine animals, always behaves in this way, boring into the walls of the intestinal canal soon after gaining admission to its host, and only emerging from its cyst when sexual maturity has been reached.

In horses and mules that have died of what was said to be an outbreak of " surrah," I have recently found myriads of these worms so encysted, but, in spite of the familiarity with the pathological appearances so gained, and a most prolonged and careful search, I have entirely failed to find any trace of such a habit in cases of anchylostomiasis. In every *post-mortem* examination, a most diligent search was made for encysted worms, not only with the naked eye, but with the simple microscope, and every point that appeared at all suspicious was carefully dissected. In addition to this, considerable areas of intestine, in the worst case I have yet met with, were examined by the method of serial sections, but without the least result. Had encysted worms been present, I am positive that they could not have been overlooked, and hence I feel some doubt as to such an occurrence being the normal habit of the worm. Another argument against it is to be found in the fact that on working out the life history of *Sclrostomum tetracanthum*, I found that its infective rhabdites were provided with a distinct and peculiar single boring tooth, and it is natural to suppose that a similar structure would be found in the *dochmius* rhabditis if it were addicted to a similar habit.

Of course, it is possible that, in the cases I have had the opportunity of examining, all the worms had passed through the stage of encystment and regained the lumen of the intestinal canal. All that can be said is that it is certain that no such condition was present in either of these cases. With one exception, however, all these *post-mortem* examinations were on cases that had been a considerable time in hospital, and so debarred from the chance of recent reinfection. The nearest approach to anything like encystment met with was in searching over a set of serial sections, some of which showed very young worms, not much larger or more developed than those found in the experiments on monkeys, imbedded in blood clot which filled up recesses between rugæ of the mucous coat. In one instance a distinct erosion of the mucous membrane was found underlying the clot. This was from the case which was only a few days in hospital before death.

The case in favour of encystment as a normal occurrence, may be summed up as follows :—

(1) In two closely-allied species, of similar habits, it has been proved to occur as a normal part of the life history.

(2) All worms found in the lumen of the intestine are almost always adult and full grown.

(3) The fact of but few observers having met with encystment is explicable by the fact that cases in which autopsis are obtained have been usually protected from re-infection for some considerable time by residence in hospital.

(4) Cases not unfrequently occur in which, after all worms living within the intestinal canal have been cleared out by vermifuges, ova re-appear in the fæces within a short time. I have met with many such cases, but they are equally explicable either by re-infection, or inefficient action of the remedy.

However, it is certainly not common in ordinary *post-mortem* examinations, and the point, which is a comparatively unimportant one, must remain open.

It is clear that, in order to observe it, one must meet with the fortunate coincidence of death from other causes, in an early case of anchylostomiasis, or one in which recent reinfection has happened; indeed it is probable that only the former class of opportunity would serve, as many clinical facts tend to show that advanced cases, from their inability to afford the requisite nourishment, do not form a favourable home for the parasites, so that it is possible that, in such cases, newly-introduced embryos may, as in monkeys, be unable to proceed with their normal metamorphoses.

On the whole, I am inclined to believe that encystment of the newly-introduced worms will prove to be a fact, and greatly regret that my scanty pathological opportunities have as yet failed to afford me the opportunity of observing it.

If, indeed, this wandering into the solid tissues of the intestinal wall be a fact, it affords a ready explanation of the occurrence of pyrexial attacks in the earlier stages of anchylostomiasis, as such a wholesale wounding of so septic a surface as that of the lumen of the bowel could hardly fail to cause more or less septic poisoning: moreover, the analogy of trichiniasis, where fever is a common accompaniment of the migration of the worms, would lead us to expect it, quite apart from theoretical considerations.

The duration of the life of the worm in its parasitic stage is a point on which, from the nature of the case, only a surmise can be formed. The failure of all experiments on animals makes the production of definite facts an impossibility. Some authors speak of a year or years, but I believe such an estimate is much beyond the truth. They have probably founded this opinion upon the well-known long duration of individual cases of the disease, but, in so doing, have quite overlooked the possibility of repeated infections. The number of available ovum germs in any given worm is, however, limited, and the rate of production very high, and as analogy would lead us to expect that the worm would die as soon as sexual activity became impossible, it can hardly be supposed that its life can extend over many months.

Another strong argument against the longevity of the worms is that the number met with, in *post-mortem* examinations in fatal cases, is often obviously inadequate to account for the mischief that has been effected, and this can only be explained on the assumption that, at some period of the case, a very much larger number were present. Owing to their home being in the duodendum, worms dying a natural death would undergo digestion in their passage through nearly the whole length of the intestinal canal, and would not appear in the dejecta. That this does not take place when they are expelled by thymol, is owing to the fact that the worms are much more often stupefied than killed outright by this drug, and often still exhibit feeble motions. Those that have been really killed are always found to be more or less digested.

From these considerations, I believe that the disease has a natural tendency to cure, but too often the damage inflicted by the worms during their life has been so serious, that the patient cannot muster sufficient strength for recovery, whether they be expelled in the course of nature or by the action of vermifuges.

VII.—Effect of various Conditions on the Life of the Free Stage.

Having thus traced what may be termed the normal life history of the *Dochmius* rhabdites, it remains to be seen how they are affected, adversely or otherwise, by external conditions.

Effect of Nourishment.—Some points in this connection have already been incidentally noticed, but as various considerations of great practical importance in devising · preventive measures are thereby elucidated, it will be necessary to go more into detail.

If the worms be supplied with but a limited amount of their natural food, as happens in the method of cultivation in fæcalized sand, already described, they pass through all the stages already described, but only a single generation of mature rhabdites appears. After this the supply of nourishment is so far exhausted, that the progeny of those which have reached maturity fail to do the like, and remain in the undifferentiated, non-sexual stage. So far, however, from shortening their lives, this deprivation of a free supply of nourishment, and the attendant arrest of development appears to greatly prolong it.

As we have seen, with a free supply of nourishment, the whole duration of life is a matter of only a few days; but if development be arrested by exhaustion of the nutritive supply, the embryos remain, without growing or changing in any way, for very long periods. What the limit of this may be, I am unable to say, because my experiments on this point are still pending. I can only affirm that, after fully six months, my cultivations seem as rich in embryos as ever. When in this state, the embryos appear to rarely, if ever, change their skins, they are sluggish, and move but little, in fact, they are in that stage which Lutz described as encapsuled. The term, however, appears alike needless and misleading, for their structure is quite unchanged, and the so-called capsule is nothing more than the ordinary chitinous investment common to all nematodes. I have dissected a great many specimens, and am convinced that nothing whatever of the nature of a capsule is present. When first transferred from damp sand to a tray of water, they are extremely sluggish, but soon begin to move about actively, and after a while they again become quiet, and often remain quite motionless for days; but the addition of some stimulating agent to the

H

water, such as a little salt or alcohol, quickly induces motion, and convinces one that they are not dead, but resting. It may be that the comparatively limited supply of air and general unnaturalness of their surroundings have something to do with this, for cultivations conducted by pouring the fæces over an enclosed patch of garden soil behave somewhat differently. Several experiments were made in this manner, four planks being nailed together so as to form a bottomless box, about 9 inches square, and sunk in the soil of the garden. Infective fæcal matter, diluted with water, so as to spread easily over the whole surface, was then poured on the little patch of ground so marked out, and the rest left to nature. Under these circumstances, judging from the normal length of life of the rhabdites and the length of time during which they continued to appear, some two or three generations may be expected, but, in the end, with the exhaustion of the supply of nourishment, the same state of things is brought about as appeared earlier in cultivations carried on within the laboratory, though of course the number of embryos left in the resting stage is much increased.

Development, however, is but arrested, and only awaits a fresh supply of nourishment to recommence. If we now pour over such a cultivation some quite uninfected fæcal matter, in a couple of days, mature rhabdites of both sexes appear in immense numbers and continue to appear as long as the fresh nutritive supply lasts, after which, the embryos, greatly increased in numbers, sink into the resting stage and may remain so, as we have seen, for months. After which the process may be repeated again, with the same result.

The following experiment will illustrate these points :—

*Experiment 24.—May 3rd.—*On visiting the Shillong dispensary I found a man who was suffering from undoubted symptoms of anchylostomiasis. He came from a village near Nongpoh, on the Gauháti road, a neighbourhood in which *kála-azár* is beginning to appear, and had never been employed on a teagarden, being in fact an ordinary villager. Dr. Campbell had already recognized the nature of the case, and had treated him once with thymol. Examination of the fæces, however, showed that there were still some worms in the intestine, as they contained a considerable number of ova. Some of the dejecta were placed in a bottle and allowed to remain undisturbed for two days.

*May 5th.—*Eggs far advanced in segmentation, but no embryos. Fæces diluted with water and poured over a patch of garden soil enclosed between four boards.

May 6th.—There is but little fæcal matter to be seen. The ground has been freely turned up by small beetles. A number of pediculi are to be seen, also one or two other insects. Specimens of these were dissected, but neither eggs nor embryos could be found within them. A number of rhabdites embryos have hatched out.

May 9th.—Well-developed rhabdites present. A setigerous annelid dissected, contained one or two embryos differing in no way from those which had not met with this fate, but, at the same time, apparently none the worse for it.

May 27th.—Mature rhabdites still present, but not in large numbers; sluggish undeveloped embryos present in large numbers.

June 5th.—Embryos plentiful, though a couple of dry days has dried the surface considerably. Still one or two mature specimens present.

July 8th.—Numerous sluggish non-sexual embryos present, but no mature rhabdites.

July 11th.—Examined with the same result, except that one or two sexually differentiated, but yet not mature, specimens were met with.

July 20th.—A quantity of fæcal matter from a healthy person carefully examined, and found free from ova, mixed with water and poured over the surface of the cultivation.

July 22nd.—All offensiveness has disappeared, principally through the agency of small caleopetra; mature rhabdites present in considerable numbers.

August 11th.—Numerous embryos in the resting stage, but no mature rhabdites.

August 12th.—Again examined with the same result.

August 14th.—Again examined with the same result. A fresh dressing of healthy fæces applied.

August 15th.—Numerous rhabdites in the stage of sexual differentiation, preceding maturity, present.

August 17th.—Mature rhabdites in large numbers, also very young ones, obviously just hatched.

August 20th.—Rhabdites in large numbers still present. The result of the two fresh supplies of nutritive matter has been to so greatly increase the number of embryos present that twenty were counted in a morsel of earth not as large as a pea.

August 28th.—No mature rhabdites, but only small embryos as yet without sexual differentiation present.

I look upon the facts established by these observations as by far the most important result of my investigation. It shows that a patch of ground on which fæcal matter has been deposited by a person suffering from the disease, may remain infected for many months, and that the accident of a fresh deposit of fæcal matter, even from a healthy person, may enormously increase its infectiveness, and indefinitely prolong it. No one who knows anything of the habits of natives will doubt for a moment that the revivication of infected spots, by a fresh supply of nourishment, must be an accident constantly happening in practice, as each family usually confines its attention to a small patch of ground, every part of which must be used as a place of deposit again and again, in the course of the year. Further, as a spot may remain infected for months after all signs of defilement have passed away, there is nothing to prevent the villagers from using the infected earth for cleansing eating-utensils, and for all the other numerous purposes to which mud and earth are turned in native domestic economy. Most important of all, it shows that there is no limit whatever to the time the infection may hang about an inhabited place, as, though it is to be supposed that a time will ultimately come when the resting embryos in my cultivations will succumb, and pay the debt of nature, yet, at any moment before this takes place, a fresh supply of fæcal matter would suffice to start it off on a fresh lease of life ; and under the conditions of life subsisting in an Assamese village, this must, as has been pointed out, occur again and again.

Effect of Supply of Oxygen.—In all stages, the embryos exhibit a strong preference for situations in which the air supply is of the freest ; if we examine a cultivation, all of them will be found quite on the surface. In the deeper portions of cultivations, whether in the laboratory or in the open, the ova either do not hatch out, or else the embryos quickly work their way to the surface. Owing to this, cultivations in the open are always much more fruitful than those undertaken in the laboratory, because, in addition to the freer supply of air, the busy operations of beetles and other insects turn up the soil into a fine light powder, and largely increases the depth to which air can penetrate, in this way, more than compensating for the nutritive matter which they consume. Owing no doubt to this cause, ova placed in water, for example in a test tube, hatch out much more slowly than those spread out on sand or garden soil, and when hatched out, the embryos fail to undergo complete development. Lively as their movements are in water, they are quite incapable of raising themselves in it, and, on account of their relatively high specific gravity, sink to the bottom with great rapidity, and remain there. In spite of this, embryos, at any rate such as have been reared in sand cultivations, will live for weeks in a considerable depth of water. A depth of a foot, however, was the greatest I have been able to arrange for conveniently, but even at this small depth, however well supplied with nourishment, though able to exist in a resting state, they will neither breed nor develop. In very shallow trays of water they will continue to develop for a short time, but, even here, soon come to a standstill. Evidently, water is not their proper habitat, which indeed is to be found only in defiled soil. Even in shallow trays, embryos in water died out altogether in from two to three months, whereas, as we have seen, the degree of longevity of embryos in sand cultivations has not yet been reached. As might be expected from the confinement of the embryos to the surface of a cultivation, burial of infected fæces will entirely prevent development, and not only so, but brings about the death of the ova so that when unearthed and exposed to the air, no development takes place.

The following experiments well illustrates this.

*Experiment 20.—May 2nd.—*A wide tube 9 inches in length was taken, and closed air-tight at one end with an India-rubber stopper. Some fæcal matter from a patient in the Shillong dispensary, which was ascertained to contain *Dochmius* ova in large numbers, was mixed with sand and water exactly as in making the usual cultivations, and placed at the bottom of the tube, *i.e.,* lying upon the stopper. Over this was placed damp sand until the tube was filled to the mouth, and a piece of coarse muslin tied over all. By this means it was easy, by removing the stopper, to examine the buried cultivation without disturbance, and with only a very short exposure to the air. At the same time, a check experiment of a cultivation of the same fæcal matter, conducted in the usual way, was made.

*May 8th.—*Check experiment swarming with embryos.

*June 3rd.—*Withdrew the stopper, and examined the buried cultivation. It exhibited not a sign of nematode life. By close examination under higher powers, some bodies, which appeared to be degenerated ova, were made out. The tube was now inverted, and the surface, which was still sufficiently damp, was left exposed to the air. The check experiment still contained abundance of embryos.

*June 12th.—*No embryos can be made out, nor did any appear at any subsequent time, although the surface was kept damp for a long time. At the date of writing, August 30th, the check experiment still contains numerous embryos.

It was found that dry sand or earth was just as effectual in preventing all development as damp sand. The practical importance of these results cannot be over-rated, as it clearly shows that burial in trenches, as practised in the usual plan of dry earth conservancy will be quite sufficient to render the infective matter absolutely innocuous, and that, once buried, there is no reason to fear any subsequent infection either through drinking-water, or by means of any other vehicle.

INFLUENCE OF MOISTURE.—A certain amount of moisture is essential to the well-being of the rhabdites. The most favourable state of things is when the soil is moist enough to hold together, but yet not sodden; extreme conditions of either dryness or moisture being unfavourable. On the other hand, although their development is thereby arrested, they can retain their vitality, in spite of drying, for considerable periods. I have by me, a cultivation which has been repeatedly allowed to become powder-dry and to remain so for weeks together, and which yet, a few hours after damping, shows abundance of lively embryos. If, however, the dessication be carried to an extreme degree, the embryos are thereby killed.

A portion of a cultivation rich in active embryos was chemically dried, without the aid of heat, by being placed under a bell-glass, along with a vessel containing concentrated sulphuric acid. After a week, the material was removed to the open air, and moistened freely with water. Many embryos could be made out, quite shrunken and showing no signs of life. After several days in water they had swollen out somewhat, and were commencing to decompose, but showed no signs of vitality.

The direct rays of the sun, if allowed to act for a considerable time seem, to have the same effect. A great proportion of my earlier observations were made on an accidental cultivation which existed outside the dead-house of the charitable dispensary at Gauháti. The piece of ground in question was a steep bank, which received the outflow of the mortuary drain. Over this drain the washing operation involved in the search for parasites was usually conducted, and accordingly the ground outside had been freely irrigated with water, carrying with it large numbers of ova. During the whole of the cold weather, the ground remained continuously moist from the heavy morning fogs, and afforded an unfailing supply of embryos, in all stages of growth. It was just at the end of the cold season that I left Gauháti, and, on my return, the bank had been subjected continuously to the action of the sun during the brief duration of the Assam dry season.

Not a trace of embryos could now be found, and, that they were killed outright, is evident from the fact that, during a recent visit, after the rains had been falling for months, the ground was still as barren of nematodes as if it had never been infected. The situation was, however, singularly exposed, and received the full glare of all the sunlight that could be got, so that the conditions were exceptional for Assam, where the dense vegetation, which always is encouraged round houses, must render a sufficently intense dessication to ensure the death of the embryos a most exceptional occurrence, and it is just in the shady bamboo, and plantain-groves about the houses that infection is commonest and most dangerous.

This observation has an important practical bearing, as it offers us a ready method of disinfecting sites, by clearing them as thoroughly as possible at the end of the cold season, and leaving the bare surface exposed to the sun throughout the dry hot weather. It shows too, that the spread of the disease in very dry climates, such as the Punjab, is extremely improbable, even should it be introduced into such localities, as there would only be a few months in the year during which infection of the soil would be possible, and the whole infected area would be quickly rendered harmless, once the short period of damp weather was past. On the other hand, the toleration exhibited by the embryos for moderate degrees of drying must be the source of an added danger in climates where dessication can rarely go to an extreme extent, for it is evident that the partially dried embryos must often form a constituent of village dust, and get carried by the agency of the wind to new sites, where chance may at any time find them the nourishment necessary to start a new focus of infection. On the whole, it seems probable that rhabditic *dochmii* possess the power of resisting dessication in as great a degree as any nematode, for they will resist ordinary drying for very prolonged periods, and I cannot recall any notice of experiments on other nematodes where they have been subjected to actual chemical drying.

INFLUENCE OF HEAT.—The extent to which the development of the embryos is influenced by the temperature of the air has already been touched on. It remains to notice the influence of higher temperatures.

A series of experiments was instituted to determine the temperature at which infected fæcal matter would be rendered sterile (as far as concerns the development of *dochmius* rhabdites). These experiments demonstrated that any temperature exceeding 140° F. is fatal alike to ova and embryos. At intermediate temperatures, such as 120°-130° development is retarded, much in the same way as by cold, but not put a stop to. A very short exposure to a temperature exceeding 140°, for example, by raising the water bath to 150°, and then removing the lamp, and allowing it to cool naturally, was quite sufficient for the purpose. It may be noted that in hot dry climates, such as the Punjab, the temperature of the surface of the soil must not unfrequently reach this limit during the day. Throughout this series of experiments, check observations to test the infectedness of the fœcal matter employed were uniformly carried out.

INFLUENCE OF CHEMICAL AGENTS.—Perchloride of mercury, thymol, carbolic acid, &c., were tried, but only proved fatal when employed in such considerable strength as to render their employment too expensive to be practicable, except in the disinfection of dejecta in hospitals, &c. Of these, perchloride of mercury is by far the most efficient, the addition of an equal bulk of 2 per thousand solution being quite sufficient to put a stop to all development of embryos. Viewing the known carelessness of sweepers and other hospital servants, I think this precaution should always be taken in hospitals where cases of anchylostomiasis are under treatment, especially as solution of this strength costs so little that there can be no objection to the adoption of the plan on the score of expense.

INFLUENCE OF LIGHT.—The more natural the surroundings of an experiment, the more rapid and abundant will be the development of the ova into embryos and mature rhabdites, and I have a general impression that the influence of light may have something to say in the markedly better results obtained in experiments conducted in the open air, over those carried out under laboratory conditions. At the same time, that complete development to the adult *rhabditic* stage and their reproduction may take place in complete darkness is a fact that has been demonstrated, during the present investigation, by a large number of experiments, so that the influence of light is, at any rate, not essential.

VIII.—THE METHOD OF INFECTION BY THE PARASITE.

Having now considered the life history of the parasite, and the circumstances that are favourable or hostile to its development, the path, or paths, by which the infection is carried to human beings, remains to be discussed.

In the first place, it must be remembered that the ova themselves are never the infecting agent. Apart from theoretical or rather comparative considerations, which are entirely against the possibility of direct infection by the ova of species possessing a free stage, the circumstance that embryos are never found in the fæces, however prolonged constipation may have been, is alone sufficient to prove this. In other words, the intestine does not offer suitable conditions for even the hatching out, let alone the development of the eggs produced by the parasite, and, as the other conditions of food and temperature are quite favourable, it may be presumed that the wanting factor is a sufficient supply of oxygen, an agent which, as we have already seen, exercises a most potent influence on all stages of its development.

It is the hatched-out embryos, and these alone, that are in a fit condition to develop into the entozoic stage of the worm, and, from the analogy of other similar parasites, it is more than probable that the embryos hatched out from parasitic eggs are innocuous, and that it is only the progeny of the free stage that is capable of infecting.

Another matter that is certain is that the embryos must gain access to their human hosts by the mouth, and so must be introduced either with the food or drink, or perhaps, in certain instances, by the fingers, when casually introduced into the mouth for other purposes than eating.

Hitherto, it has always been assumed that the parasite gained access to the intestinal canal through the agency of drinking-water, but this opinion has been based rather upon a priori considerations than upon the observation of actual facts, and, though I have no doubt it occasionally occurs, I am personally inclined to believe that drinking-water forms a quite exceptional vehicle of infection. In the first place, out of the very large number of examinations of drinking-water from infected places, in collecting which I in many instances was at pains to stir up the mud, so as to favour the discovery of embryos as much as possible, I have only once met with any sign of *Dochmius* embryos, and even in this case, as the specimen was a much-decomposed female, and not one of the very characteristic males, it is quite possible that it may have belonged to a harmless and non-parasitic species. I can only say that it was a rhabdites, agreeing in general size and appearance with those of our species. Now, were drinking water the usual vehicle, it is probable that it would have been met with often, in large numbers, and not an isolated specimen, once only.

Again, the physical peculiarities of the embryos are entirely unfavourable to such an occurrence. Supposing them, as indeed the ynot unfrequently must do, to gain access to a well, they would at once sink to the bottom of it, and become entangled in the mud, and, when we remember what clear water may be commonly drawn from wells very muddy at the bottom, it must be seen that, unless they happen to be taken during their very rapid passage to the bottom, their mixture with water actually taken for drinking must, in the case of wells, be a most exceptional occurrence. Infection by water taken from running streams must be still more exceptional, for the immense dilution that infected drainage must at once undergo, must render the chances against even a single embryo being contained in the few ounces of water drunk during the day very small, for, prolific as these nematodes are, the numbers produced are in no way comparable to those of bacteria, such as those of typhoid fever, where the immense number and infinite smallness of the infective organisms renders the infection of large volumes of water not only possible, but common. In the case of large rivers, such as the Brahmaputra, water carriage of infection is well nigh inconceivable, for it must be remembered that the introduction of single embryos would do no appreciable harm, and that, unlike the bacteria, they are quite unable to breed there like within the human subject, so that every worm found parasitic must have been separately introduced from without. Such being the case, how can drinking-water have been the vehicle of infection in the already cited instance of the village of Paru, whose only possible source of drinking-water is the river Brahmaputra, the main current of which sweeps by, within a few yards of its houses. On the other hand, *bhils* offer much more favourable opportunities for the conveyance of contagion, but, even here, the natural habit of avoiding scooping up more of the mud than can be helped must operate to render its carriage in water taken for drinking very rare.

The habit which natives have of washing out their mouths with water which they have just stirred up by bathing in it must, however, offer exceptional opportunities for infection by the agency of water, especially as they nearly always ease themselves on the very brink of the water in which they propose to bathe. In pools employed for this purpose, I have repeatedly met with undoubted *Dochmius* embryos, and have little doubt that this habit is an occasional source of infection, although the people always stated that they did not use the water for drinking. It is improbable, however, that this will account for more than a very small proportion of cases, and, in numbers of villages, although very badly infected, there is no opportunity for its occurrence this way. For example, in Paru the people bathe in the river, the rapid current of which would instantly carry away embryos that might get into it. If water, then, be, at best, a very exceptional vehicle of infection, what, it may be asked, is the usual way in which the disease is contracted? To answer this question, we should first ascertain in what situations infective embryos are to be found in the largest numbers. These situations, I must most emphatically repeat, are neither wells nor pools, nor collections of water of any kind, situations which I have, moreover, experimentally proved to be entirely unfavourable to the life of the embryos, but the surface of the soil near dwellings, which, as we have seen, offers by far the most favourable situation for their development.

In most parts of India, the villagers, however dirty their habits, make a custom of going to some little distance from their houses to relieve nature. As they generally choose the banks of the nearest *nullah* and so render inevitable the defilement of their water-supply, this habit is, from a general sanitary point of view, even worse than that of the Assamese. One naturally expected that in this matter their habits would resemble those of other Indians, but I was surprised to find, from careful personal observation, that they rarely trouble themselves to go more than a few yards from their own doors. It is not uncommon to find the younger men conform to the Indian habit, but old men, women, and children, and especially the sick, will not move more than a stone's-throw from their huts. After all, it is but natural that a man scarcely able to move from the weakness brought about by anchylostomiasis, and often tormented by diarrhœa, should not trouble himself to tax his strength by going to any considerable distance, especially as his ideas of decency in this matter are, at the best, quite rudimentary. Owing to such habits, a broad circle of ground round the hut of a patient affected with anchylostomiasis becomes little better than a large cultivation ground for the embryos, and, in such situations, there is never any difficulty in finding abundance of them. Assam is essentially a muddy climate: even in the cold weather, the daily morning fogs keep the surface of the soil continually moist, and, when it is much trodden upon more or less muddy, and it is only during the short dry season that the surface becomes absolutely dry, and, even then, the soil will be found to be damp at a very little depth. Hence, in entering his hut, and in moving about amongst his neighbours, the inhabitant of an infected village must always be getting his feet covered with mud, which is absolutely teeming with infective embryos. When we remember that a bit of mud no larger than a pea may contain scores of them, it is easy to see what multitudes must be introduced into the house each time the occupants enter it. Once within the house, there are many ways in which the embryos may gain access to the intestinal canal of the inhabitants. For example, food is always eaten squatting on the ground, with, at most, a bit of seldom-cleaned matting as at once table and table-cloth. Under such circumstances cleanly eating is impossible, and a certain amount of the dirt of the floor is pretty certain to get into the dish at each meal. Not only so, but, in the position natives sit when eating, their hands necessarily often come into contact with their feet, and, as they have no other forks than their fingers, these latter must often convey into their mouths dirt from their feet or the floor. Again, in course of time, the infected mud, which must necessarily be introduced into the house, will become house-dust. Now the very moderately intense drying that would take place in such a house, would be quite insufficient to kill the embryos, and being, when dry, very light, they must be raised into the air with the other constituents of the house-dust, and descend on food and upon cooking utensils, and it is quite possible that some may be drawn into the nasal passages in respiration, and so gain access to the stomach together with the nasal mucous, which is constantly flowing, through the posterior nares, into the pharynx. There are other native habits which also greatly favour infection. One of these is that of cleaning brass eating utensils with earth taken up anywhere, and I have actually seen a woman cleaning her platters with earth which I ascertained to be infected, and, as an instance of their utter callousness to the most ordinary considerations of niceness, I may mention that in this very instance I saw her husband calmly easing himself, on the same patch of ground not many yards off. Another habit favouring infection is that of smearing the walls and floors with mud mixed with cowdung, in the preparation of which compost infected earth is often used. In its application no other trowels than the hands are used, and the users are, we may be sure, none too nice in cleansing them before employing them to handle their food. In fact, the habits of the Assamese could not better favour infection had they been specially devised with a view to the purpose, and the only wonder is, not that so many are infected, but that any escape. Nor are such habits confined to the poor and wretched ; in fact the advantages enjoyed by the educated clerk in an English office consist more in better clothing and food than in any essential difference in the sanitary tendencies of his domestic habits, so that they in no way share the immunity from attack that is enjoyed by Europeans. The longer the disease remains in a village, the wider will the area of ground become, the surface of which will teem with thousands of infective embryos.

As far as my experiments on this point go, they tend to show that the embryos do not spread very widely around the centre on which the ova were deposited. They are too much entangled in the surface mud to be spread very widely by ordinary rain. Still, in time of flood, especially where there exists a current of any strength, the embryos must often be carried to considerable distances, and, wherever they are deposited, they will lie in the resting stage for months, only awaiting the accidental deposit of a fresh supply of fæcal matter to breed and multiply, and start a fresh centre of infection. In this way, large portions of the area of *kála-azár* villages may come into such a state that the accidental introduction of a small portion of the surface soil into the intestinal canal will be sufficient to produce the disease in a new host. Unobservant as the peasantry are, they have at least recognized the persistent way in which the disease clings to any place in which it has once established itself, and know well that no measure of prevention is so effectual as the entire abandonment of such sites.

From what has already been said, it will be seen that, as a rule, infection must be a very gradual matter. Cases may, and no doubt do, occur in which the accidental swallowing of a large dose of infective matter may suffice to introduce a number of embryos large enough to cause serious mischief, and, indeed, some cases of an acute character can hardly be accounted for in any other way ; but usually the process must be a very gradual one, and far from being a matter of a single infection, the process more probably is a matter of almost daily small increases from the beginning to the end of the case, during which I am inclined to believe that often several successive sets of parasites may come and go before the fatal termination. It is owing to this, that treatment out of hospital is such a discouraging, and indeed hopeless, matter, for, owing to the necessarily constant recurrence of infection, only a temporary improvement can be expected. The ground round the hut that has been occupied by a case of anchylostomiasis must soon get in so highly infective a condition that, if the patient is to return to it, treatment can be little better than a waste of drugs.

Probably in no two cases is the rate of infection uniform or similar, and the spread from man to man in a village must necessarily be equally irregular and uncertain for the individual, though inevitable for a considerable proportion of a village population. The rate and method of spread are in fact just what we find in *kála-azár*, and are necessarily entirely different from the phenomena to be observed in directly infective diseases, such as the specific fevers. In anchylostomiasis, the infection is indirect, and mere contact and association with a patient are quite incompetent to transmit the disease, whereas, in ordinary communicable diseases the infection is direct, and there is usually no difficulty in tracing its source. It was undoubtedly with the phenomena of direct contagia, such as these, alone in their minds, that previous observers of *kála-azár* have so confidently stated that there is no evidence in favour of its contagiousness, and, so limited, their deduction was perfectly correct. In reality, however, anchylostomiasis, for people in a low grade of civilization, is one of the most infectious of all maladies, taking the word in its broad sense of communicability, and the method of its spread coincides exactly with the facts observed in connection with the spread of *kála-azár*.

On the other hand, the danger of infection is almost *nil* for Europeans, even in India. The mere fact of eating at a table, with knives and forks instead of fingers, is alone almost sufficient to secure immunity, and in even the most backward European towns, the system of conservancy, however badly it might be conducted, would suffice to prevent all chance of contracting the disease. Were the whole of the affected population in Assam suddenly transferred to London to-morrow, not a single case of infection could possibly occur, as long as the new-comers were forced to live under the same conditions as Londoners.

There is, however, one danger of infection for Europeans in India, about which a word of warning may not be out of place. I refer to milk supply. In Gauháti, for example, the danger is a very real one. The greater part of the milk consumed by the residents is brought into the town from Paru, and other villages, where *kála-azár* is rife. It is conveyed in galvanized iron buckets, and any morning the milkmen may be met coming in, each with a couple of buckets slung on the ends of a bamboo. To prevent the milk splashing and being lost, they place in each bucket a wisp of reeds or grass. Now, it is perfectly obvious that the reeds may have been taken from a *bhíl* swarming with embryos, and it is even more likely that the udders of the cows, and the hands of the milkers may be fouled with mud so infected, and though I have met with no case where any suspicion of infection of a European has occurred, it behoves all residents of such neighbourhoods to be careful that all milk used in their houses is thoroughly boiled,—a precaution which will certainly render safe any specimen of milk, however badly it might be infected.

IX.—Remarks on the Pathology, Diagnosis, and Treatment, &c., of Anchylostomiasis.

To attempt any exhaustive treatment of this subject would involve writing a treatise of nearly the bulk of the entire present report. Moreover; the task has already been too well accomplished to make it either necessary or desirable that I should do so, and all that is here attempted is to emphasize some old points that appear of special importance, and to notice a few new ones that have cropped up during the present investigation.

The importance of an early diagnosis in cases of anchylostomiasis cannot be overrated. If we wait till matters are gone far enough for the patient to be unfit for work, the chances are that the time has passed during which he might have been saved. When first it was discovered that *kála-azár* was actually anchylostomiasis, but having no previous practical experience in treating the disease, I naturally hoped that the expulsion of the parasites would be sufficient to initiate a cure, and the rest would be merely a matter of waiting until a sufficiency of food had been assimilated to make up for what had been lost. This *a priori* notion would doubtless have proved correct enough had the patients retained the power of assimilation; but a further experience of the disease has demonstrated the fact that the mere loss of nutritive matter required for the support of the parasites is a small and quite unimportant factor in the causation of the fatal symptoms produced by them. Various calculations have been made as to the amount of blood actually withdrawn by the parasites, and these, though differing pretty widely, agree in making the amount too small to be likely to do more than seriously weaken an adult. The real damage is mainly caused by the destruction of the digestive powers. A large share in the causation of this diminished power of assimilation, is probably to be accounted for by the constant recurrence of great numbers of small traumatic lesions of the mucous membrane. Should these lesions include boring, and temporary encystment, the matter is the more easily understood; but the bites themselves, in spite of their small area, can account for much.

Microscopical examination has shown me that the gnawing of the parasite produces an erosion which extends to the submucosa. Now, we know that, once damage of an epi or endothelial surface has gone to this extent, repair takes place by the formation of a scar, and not by reproduction of either true skin or secreting mucous membrane, so that each bite must leave a small area incapable· of ever again performing its ·natural functions. The number of bites always largely exceeds that of the parasites, so that each probably inflicts several · small injuries in the day, and when this is multiplied by a large number of parasites, acting through a long period, the total area so affected may become very considerable,. especially as the duodenum and upper part of the jejunum are one of the most important portions of the intestinal tract. Again, in the examination of advanced cases, the stomach is always seriously affected·; it usually shows scattered petechiæ and a peculiar mammilated, leathery appearance of a dull slatey red hue, which is most often best marked along the greater curvature, and which reminds one, more than anything else, of the grained appearance of morocco leather. It is undoubtedly this change which is referred to in Mr. McNaught's accounts of *post-mortem* examinations of *kála-azár* cases. This mammilated appearance is so peculiar that, on first seeing it, I was convinced that it must be due to small cysts in the mucous membrane, and that the encysted parasite would readily be found in them. Careful microscopical examination, however, failed to show any sign of the parasite, but only an elevation of the mucous membrane caused by inflammatory infiltration of the submucous tissue, and often enclosing a small old blood clot. It is very possible, however, that they may be the remains of cysts, which once contained parasites, which have escaped in the due course of their development. Their close resemblance to the changes found in the stomachs of equine animals affected with *Sclerostomum teracanthun*, where there is no doubt about encystment, strongly supports this view.

The ultimate result is a chronic gastritis, with thickening over certain portions of the mucous membrane and atrophic changes elsewhere; the very process of repair leading ultimately to wasting and functional uselessness of the organ. The changes in the intestine are very similar, though large thickened areas are rarely met with. The principal organs of digestion having been reduced to this condition, proper assimilation becomes impossible, and a condition of slow starvation follows, under which the losses of blood, small in themselves, rapidly produce the perniciously anæmic condition with which we are so familiar.

Apart from this, it is well known that starvation, or any condition equivalent to it, may, in itself produce, changes in the

I

assimilative mucosa, which, though difficult of exact anatomical definition, are none the less quite fatal to the function of these organs. The most prominent symptom 'of this is usually a looseness of the bowels, the well-known "famine diarrhœa," so often noticed in reports referring to the great Madras famine, though constipation or an alternation of these conditions were sometimes met with. In such cases, recovery was practically unknown, and that in spite of the most careful feeding, which the digestive organs having almost lost their assimilative powers, had become practically useless. Now, the condition into which patients are brought by anchylostomiasis closely resembles that of starvation, and no one can look at old photographs of the victims of the famine, without being struck with the fact that a similar series taken from a *kála-azár* village would be quite indistinguishable by the closest scrutiny. Moreover, the symptoms of the fatal dyspepsia are closely similar in both, though modified by the circumstance that, in anchylostomiasis, the bowel, in place of being empty, habitually contains plenty of food. It is probably owing to this that constipation is a much commoner symptom in the disease we are considering than in the dyspepsia of famine, the watery flow being mechanically stopped by the mass of undigested matter encumbering the lower bowel, until re-absorbed, and which it has not the strength to expel. In spite of this, attacks of diarrhœa of long duration are an almost constant symptom of anchylostomiasis, but usually these attacks alternate with shorter periods of constipation.

I am anxious to give great prominence to the primary importance of dyspepsia in the production of the cachexia of anchylostomiasis, because I meet with many who, from failure to appreciate this fact, attach an undue importance to the failure of thymol to effect a cure in advanced cases, arguing that, because the expulsion of the parasites fails to cure the disease, they are therefore not its cause, and that there must have been some antecedent cause at work, whereby the patient's economy is rendered a suitable home for the parasite. Now, this is an argument that, at one time or another, has been raised, and abandoned in the case of every parasite that troubles the human body, and its mere production argues an entire ignorance of the broad general laws of parasitism. The healthier the host the better he serves the purpose of his unwelcome guest, and any falling off in his health must necessarily be prejudicial to the welfare of the parasite, as, if the loss of health be progressive, the ultimate result must be its destruction. In

connection with this Cobbold* remarks—" Most people, not excluding even the votaries of the healing art, following tradition,
regard the internal parasites or entozoa as creatures either
directly resulting from certain diseased conditions of their hosts,
or as organisms which would not have existed if their bearers
has been perfectly healthy. Nothing can be more absurd.
Such a conclusion is utterly at variance with all logical deduction
from known facts. It is, however, quite on a par with multitudes of other popular delusions, which, in spite of the advance of
science, will probably never become wholly eradicated from the
public mind. People who hold these notions cannot or do not
desire to reject a view, which has for them a dominating power
almost equal to that of any known religious dogma." Yet these
same people, who would at once recognize the absurdity of insist-
ing on the necessity of a predisposition, to permit of the action of
the deprivation of food, in the production of starvation, still
would assume such a necessity to account for what is practically the same state when brought about by the action of
parasites. The fact is that fallacies of this description die hard.

. Dr. Kynsey† divides the symptoms of anchylostomiasis into
3 stages,—(1) the dyspeptic ; (2) the anœmic ; (3) the dropsical—
and though the arbitrary division of the gradual processes
of disease into distinct stages is rather to be deprecated, his
classification recognizes the truth of the primary importance of
the dyspepsia, and the manner in which it actually underlies
the anœmia. Its disadvantage is that it does not sufficiently
emphasize the fact that dyspepsia, besides being the earliest
symptom, is continuous throughout the case, that, far from
diminishing when the anœmic stage is reached, it increases
in severity with the progress of the case, from first to last, and
that it is largely the cause of the anæmia, and not entirely
its effect, though the actual loss of blood caused by the
depredations of the worms has also an important share in
the resulting bloodlessness. Practically speaking, however, we
rarely, if ever, get the opportunity of recognizing the disease
previously to the supervention of anœmia, at any rate in
Assam, where cases will, as a rule, run into the dropsical stage
before they apply for medical assistance, and rarely indeed
seem to consider themselves as ill, until matters are very far
advanced. When once the anæmia has declared itself, however, the diagnosis, after a little practice, becomes very easy.

---

* " Parasites: a Treatise on the Entozoa of Man and Animals." By
T. Spencer Cobbold, m.d., f.r.c.s. J. & A. Churchill, London, 1879, page 1.
† *Op. Cit.*, page 5.

The peculiar character of this anæmia and the way in which it may be distinguished from other cachexiæ have already been described on a previous page, and so need not be repeated here. Even laymen soon become expert in recognizing the condition, and I was much struck with the way in which a gentleman managing a tea-garden in the Lakhimpur district, while walking over his estate, picked out early cases at a glance, and at once sent them off to hospital for treatment. The coolies, I may remark, seemed in no way to resent being treated in this way, and the undoubted success that has attended this gentleman's attempts to cope with the disease, serves as a ready answer to the objection that always meets one when systematic medical inspection and treatment of early cases is proposed. Nearly always the reply is that the coolies would not submit to such treatment, and would desert if one attempted to deal with the disease in this way; but here, as in the case of conservancy, the objection is purely *a priori*, and the objectors have never troubled themselves to test the soundness of their theory by giving the system a practical trial. But, even supposing the objection a real one, though I doubt if it has the least foundation in fact, it would be surely better to risk the chance of the loss of the services of a few men by desertion, than to face the certainty of the entire loss of a large number of men by death, and the diminution of working power of perhaps half the remainder. In the instance, however, which I have mentioned, I was assured that the popularity of the garden had in no way suffered, and that, on the contrary, the coolies seemed rather to appreciate the care taken of them.

Once the anæmic stage is fairly entered, there is no difficulty in making a diagnosis, which is certain enough for all practical purposes, but in very early cases some doubt may be felt, and in such cases, before submitting the patient to the ordeal of treatment with thymol, it is always well for the medical attendant to make sure of his diagnosis by microscopical examination of the dejecta. The examination is a very easy one, and requires neither the use of high magnification, nor anything more than a most rudimentary acquaintance with microscopy. The easiest plan is to dip the end of a strip of bamboo into the specimen to be examined, and with it to smear a minute quantity on an ordinary glass slide. A drop of water is then placed on the smear and a thin cover applied. By gently moving the cover to and fro on the slide, the water and the material for examination become intimately mixed together, and the slide is then placed on the stage of the microscope and systematically examined. Care must be taken that only a very small quantity is placed on the slide, as recognition of the ova is almost impossible in a thick layer. A better plan is to place a small quantity (a mass about the size of a pea) of the material in the bottom of a test tube which has been provided with a good cork, about a couple of inches of 1-20 solution of carbolic acid is then added, and a single drop of some analine dye, such as magenta. The cork is then inserted, and the contents well shaken and allowed to stand for a few minutes. A pipette is then passed down to the sediment, but not quite to the bottom, and a small quantity of the sediment allowed to run in and transferred thence to a slide. The use of the carbolic acid does away the small disagreeables of the examination, and the dye is useful in distinguishing the ova from the other bodies. They themselves are little if at all stained, except by a much more prolonged action of the dye, but nearly everything else found in dejecta takes it up greedily, so that the ova are the more readily made out by the contrast of remaining unstained.

The points of distinction between the ova of *Dochmius duodenalis* and those of other human nematodes have already been described, and the only other ova likely to be met with are those of tape-worms and flukes. All of these, however, want the clear belt which surrounds the yelk. Moreover, the ova of all human flukes are much more segmented, as also is that of *Bothriocephalus latus*, while those of *Tœnia saginata* and *T. solium* are smaller and quite spherical, instead of ovate. A much more frequent source of difficulty, however, are *coccidia*, a species sporozoon, which is very commonly found in immense numbers in the dejecta of patients affected with anchylostomiasis. Their occurrence is not constant, but is so common that they are to be found in the larger proportion of cases. In healthy subjects, I have not, as yet, met with them, and the instances where they have been present in the largest numbers have usually been advanced cases. What their connection with the disease, if any, may be, is difficult to say, but they certainly form no essential part of it, as they may be quite absent in even the worst cases.

The number in which they may be present is astonishing, and, estimated in the same way as was done in the case of ova, many millions may be passed in the day, so that, even in the diluted material, half-a-dozen may be found in each microscopic field. They vary a great deal in size, but are generally very much larger than the ova of *Dochmius*, averaging about

$\frac{1}{150}$ inch in diameter, so that it is only in the case of small specimens that any doubt could arise. Besides varying much in size, they are by no means uniform in shape, which, while occasionally regular, is more often unsymmetrical, differing in fact from ova much in the same way that potatoes do from eggs. Each coccidium contains from 4 to 12 or more *psorosperms*, and, in small specimens, where the outline chances to be oval, the resemblance to *Dochmius* ova is very close. As, however, they rapidly become intensely stained in even very dilute solutions of analine dyes, no difficulty will be experienced in distinguishing them if the second plan of examination be employed.

The occurrence of organisms of this class in human alvine discharges has been previously observed by Kjellberg and Eimer,[*] and it is doubtful whether they should be considered identical with *C. oviforme*, found in the liver and intestinal canal of rabbits, or not. These organisms sometimes cause fatal illness in rabbits, and Leuckart[†] gives a fatal case in man, reported by Gubler in Paris. The account of the case, however, reads suspiciously like one of anchylostomiasis, and it is possible that the *psorosperm* nodules found in the liver in this case may have been but an accidental complication, especially as the occurrence of such nodules in the tissues of animals is not inconsistent with health. The date of the report (1858) is prior to the St. Gothard tunnel outbreak of anchylostomiasis, which first brought anchylostomiasis, as an European disease into prominent notice, so that the presence of anchylostomes would have been unlikely to attract notice, even if present. The curious point is that, though specially sought for, no *psorosperm* nodules could be found in either the liver or bile ducts in the cases I have examined, but it must be remembered, however, that a considerable number might be present, and yet escape notice. Another curious circumstance is that these organisms may be associated with other epidemic diseases, for example Cobbold found that they were almost universally present, in enormous numbers, in the flesh of animals which had died of rinderpest,[‡] and Leuckart found them in the intestine of trichnized dogs, but neither author attributes any pathological importance to their presence in either of these cases, and, in the present instance, I hesitate to consider their

---

[*] Verchow's Archiv, Pathol. Anat, Bel. XVIII., page 527, 1860 ; *vide* also E. Hoyle's translation of Leuckart's " Parasites of Man." Young J. Pentland. Edinburgh, page 221.

[†] *Loc. cit.*, page 322.

[‡] Cobbold : " Parasites." J. A. Churchill, London, 1879, page 278.

appearance as more than a coincidence, though, when present, their attacks on the epithelia of the intestine can hardly fail to aggravate the existing dyspepsia. However, whatever their significance, they are extremely common in cases of anchylostomiasis, whether under the name of *kála-azár* in villages, or of *beri-beri* in tea-gardens, so much so, that, at one stage of the investigation, before I had satisfied myself of the wide diffusion of anchylostomiasis, and its adequacy to account for all the facts of the case, I was inclined to suspect that they might have some casual connection with *kála-azár*. The coincidence, however, is too common to be purely accidental, and it is possible, though I have no facts to support the surmise, that the germs may be introduced into man through the agency of the embryonic anchylostomes, and that their frequency in man may depend on a corresponding prevalence of psorosperm parasitic disease in the free stage, anchylostomes, in which case, one stage of the life of the psorospermia would be passed in the Rhabdites and another in the intestine of man. If this be the case, they must be a distict species from liver coccidia, and their non-occurrence in that organ is explained. It will be seen, therefore, that, though their presence forms a possible source of fallacy in the microscopical examination of dejecta for ova, their frequent association with anchylostomiasis renders it highly probable that when they are present, ova will be found also.

In any case where, in spite of suspicious symptoms, ova cannot be discovered, the test of cultivation described on page 72 should be resorted to, and I do not consider that the absence of the parasites can be considered as proved until resort to this method has been adopted.

The utterly insufficient grounds on which the absence of the parasites has been assumed by various observers has already been adverted to, and the commonest of these is the non-discovery of the worms after the exhibition of thymol. In most cases the examination has been nothing more than a casual glance at the dejecta *en masse*. I can only say that, however many may be present, it is in the last degree improbable that any will be discovered by so casual a method of examination, as their colour and small size makes detection in this manner almost impossible. The only safe plan is to place the mass on a loose surface of muslin, and to let water run on it for a considerable time (about an hour). At the end of this, a small quantity of quite inoffensive material, consisting of the worms, and larger particles of undigested matter will be left, but, even now, owing to the large proportion of the food that passes through the bowel unaltered, considerable numbers may be present, and yet not be visible except under special circumstances, and to ensure their detection, the mass should be transferred to a shallow glass vessel placed on a 'dark ground, and then carefully searched over, by slowly stirring it, so as to successively pass the whole mass in a thin layer over a portion of the dark ground. Such a proceeding necessarily involves an expenditure of time and trouble rarely at the disposal of medical men engaged in their usual work, and what I deprecate is, not that the trouble has not been taken to examine thoroughly, but that the most casual examination is made the basis of most positive assertions as to the absence of the parasite. It is just because its discovery is no easy matter that its presence has so long been overlooked. Only once have I succeeded in discovering the worm by mere casual examination, and that in a case where they were exceptionally numerous.

From what has been said, the immense importance of making a diagnosis early in the case will be at once appreciated. When circumstances will permit of such a course of action, we should never wait for the patient to seek medical aid, but should examine and make our diagnosis before he confesses himself to be ill. In ordinary practice such a plan is unfortunately impracticable, but on tea-gardens it should be quite easy to carry out. I feel convinced that the only thorough way of eradicating the disease from these establishments is by adopting the plan that has proved so successful in dealing with scurvy in military and prison practice. Just as is the case with anchylostomiasis, scorbutic mischief may go very far indeed before the patient is so ill as to be compelled to seek medical aid, and it has long been recognized that the only safe method of dealing with the disease is the early detection of cases by periodical systematic medical inspections. Luckily, just again, as is the case with scurvy, a diagnosis, good enough for all practical purposes, can be made at a glance, so that, if the medical attendant time his visit at the usual weekly muster, it need not prolong that process, as he will be able to make his examination of the men in much less time than it will take the manager to pay them.

• As the disease is essentially a chronic one, there would be no need for making such an examination every week, and once a month would probably be sufficient. All cases detected should at once be sent to hospital, and where the diagnosis is doubtful, it might be supplemented by microscopical examination, though this will rarely be required, and in all cases, where the diagnosis appears sure, the men should at once be treated with thymol. Taken at this stage, before the patient's digestive powers have, become thoroughly undermined, the treatment need only involve two or three days' absence from work, whereas, if the disease have once gained a fair hold, a prolonged convalescence is inevitable, and may require months to complete, during which the labourer is useless to his employer, and necessarily becomes too poor to supply himself with the liberal allowance of nourishing food indispensable to his recovery. Hence, I repeat, every effort should be made to expel the worms while the patient is still capable of digesting his ordinary food. Another advantage is, that the administration of thymol to a man whose strength is but little, if at all, diminished, is a trifling matter, while, if given to advanced cases it may even accelerate the fatal termination of the malady. By acting in this way, too, we put a great check upon the spread of the disease, for it is just these early cases, who are still able to move about freely, that spread the ova of the parasites far and wide over the country, and every worm killed means the destruction, or rather the prevention of the birth, of thousands of the infective embryos. Moreover, from all I have seen and all I can gather from gentlemen engaged in tea-garden practice, who have had immense experience in the treatment of the disease, there can be no doubt that the treatment of advanced cases is entirely unsatisfactory, and that we can rarely hope for a cure when once the fatal destruction of the digestive powers has been allowed to progress to any great extent, and hence every effort should be concentrated on the detection of cases while still in their earliest stage.

There is one symptom which has everywhere attracted considerable notice in anchylostomiasis, which requires some notice here. I refer to geophagy, or earth-eating. It must not be supposed that this symptom is pathognomonic of the particular parasite with which we are at present concerned. It is a common symptom of helminthiasis of all kinds, and is not even peculiar to man. The same habit has been noticed in elephants affected with flukes (*masuri*) as well as in horses affected with strongyles and other parasites.

It is difficult to see why such a perverted appetite should be caused by the presence of entozoa, especially as it is by no means always traceable to short-commons, combined with increased appetite ; as the victims of the habit will eat earth, even when plenty of proper food is easily obtainable, and even when the appetite for natural articles of diet is but small. As is well known, a similar habit is common in pregnant women, and here can only be owing to a sort of moral reflex to the stimulation of the uterine afferent nerves, though why the impulse should pass on to the higher centres, instead of manifesting itself in the usual way in a current propagated along the corresponding efferent fibre is a matter on which we can form no guess. Probably, the phenomenon is here owing to a reflex of a similar kind, initiated by stimulation of the intestinal afferent nerves. However that may be, the habit is very common among tea-garden coolies, and also in *kála-azár*-stricken villages, the inhabitants of which recognize it as one of the most characteristic symptoms of *kála-azár*. The coolies do not seem at all particular as to the kind of earth used, and a specimen, recently received from Chiknagool, contained a quantity of gravel, as well as some remarkably rough pebbles of coarse sandstone, and advanced cases, who are too weak to walk easily, will devour the mud from the floors and walls of the hut. The Assamese villagers seem to be more particular, and prefer a smooth clayey soil, which they regularly bake into a sort of biscuit, having a peculiar and extremely disagreeable smell, which they call *chikur*. The cooking seems to be rather of the nature of drying than baking, and I should doubt if the whole mass is constantly raised to 140°F. This peculiar habit was first brought to my notice by my Assamese Hospital-Assistant in one of my visits to Paru, where he got me some of these earthen biscuits from a house where *kála-azár* was very bad. I gather from him that certain medicinal powers were attributed to them, but he seemed to think this was rather an excuse for eating them than otherwise, and that, as he expressed it, they look upon them as " a sort of luxury." It is perfectly obvious that the existence of such a habit offers great facilities for re-infection by the ingestion of quantities of earth which may contain infective embryos, and such an occurrence is especially likely to occur in cases that are too far gone to move far from their beds in search of this extraordinary provender. As is well known, this habit is to be met with, in isolated instances, all over India, and I cannot help suspecting that microscopical examination of the dejecta in such cases would lead to unexpected results. When present, the desire seems almost uncontrollable. In a case which came under my care at Hoshangabad, the patient, an inmate of the Society of Friends' Orphanage there, and particularly well fed and cared for, was bedridden from hip-joint disease. In spite of all entreaties and remonstrance, she insisted on picking out and eating plaster from the floor of the ward. Ultimately, the lady in charge of the case took to tying her hands to the bed, and then, in spite of this and a long splint and weight, she would contrive to wriggle the upper part of her body off the bed and to grub up the plaster with her lips and teeth, whenever she was left alone for a few moments. My attention had not then been drawn to the connection between earth-eating and helminthiasis; but, looking back upon the case, I now feel little doubt that the presence of some parasite was the cause of this extraordinary longing. The only way of dealing with it is obviously by the expulsion of the parasites. Its existence is an added danger to those already caused by parasites, as death has been known to be caused by exceptionally large meals of this unnatural kind, and I can recall a case where this actually took place, in my own practice, in the case of a grass-cutter, while I was employed on regimental duty in the Punjab. This man had swallowed several pounds of earth, but even when the habit exists to but a small extent, it must greatly aggravate the dyspepsia which forms so important a feature in cases of anchylostomiasis. The difficulty of dealing with it, in actual practice, may be judged from the case already described.

In the matter of treatment, I have little to add to what has already been made public by previous writers on the subject. There can be no doubt as to the efficiency of thymol in expelling the parasites, but I am anxious to emphasize the necessity of adhering to the dosage* recommended in the Sanitary Commissioner's circular on the subject. I have found a great deal of variation in the practice of different people in this matter, with, as a rule, only the most vague general opinions assigned as the reason for so doing. For example, I found that the Assistant-Surgeon at Dibru-garh was employing doses of only a few grains, and, on closely questioning him, the only justification he could urge for the practice was, that he thought the patients were improved, but that he had never attempted to ascertain whether the worms were expelled or not, by even the most casual inspection of the dejecta. Probably, his real reason was the tendency to hoard up "Europe medicines," which

---

* *See* paragraph 49, Annual Sanitary Report of Assam for 1888. *See* also paragraph 48, lines 8-13, Assam Annual Sanitary Report for 1889, which recommends thymol first, and then purgatives.

amounts almost to a mania in many medical subordinates. As has already been remarked, my clinical opportunities have been but scanty, but I have tried these small doses in several cases, and have never found them expel a single worm, and, moreover, I treated some of the cases that had already been treated with these small doses in the Dibrugarh Charitable Dispensary, with the ordinary doses, and expelled the worms in such considerable numbers that it was perfectly clear that, whatever effect the small dose may have had, it was entirely inadequate, and for practical purposes it may be taken as certain that the employment of the drug in such small doses as 3 to 4 grains for adults is nothing better than a purposeless waste of an expensive drug.

The smallest dose with which I ever succeeded in getting any effect whatever in an adult was ten grains thrice repeated, but the dose was clearly inadequate, as many worms were left behind. The usual routine dose of 30 grains thrice repeated very commonly expels every single worm, as has been ascertained in several cases where the subsequent death of the patient enabled the matter to be established beyond doubt. On the other hand, even with this dose, I have often found considerable numbers of the parasites left behind, as evidenced by the uninterrupted continuance of ova in the dejecta. However, by employing a different plan of exhibiting the drug, I have several times obtained excellent results with thrice repeated twenty-grain doses, and am inclined to think that a great deal depends on the manner in which it is administered. Thymol is a body very little soluble in water, and its action on the parasites appears to be, by contact with the solid material, little or none of it appears to be absorbed into the circulation, under ordinary circumstances, so that by far the greater part of the drug passes unaltered with the dejecta. Owing to this, it is quite evident that a dose of 20 grains, carefully triturated, may be far more effectual than one of 30, which has been but coarsely powdered. Unfortunately, thymol is anything but easy to powder, and to do so thoroughly involves a great deal of personal discomfort, in the way of sneezing and lachrymation, to the compounder, so that, with the very inferior agency we are able to employ in such work in India, it is by no means surprising that the job is often very badly done, and, as a matter of fact, I have often, in searching for the worms, recovered a large portion of the dose from the dejecta in comparatively large crystals, often reaching the size of a No. 8 shot. Putting aside then such absurd doses as 3 or

4 grains, the effect of a dose will depend much more on the fineness of the particles of the drug than on the actual weight employed, and so, as at present administered, comes to depend mainly on the endurance of the compounder, and the vigour of his arm. There is, however, not the least reason why this should be the case, as there are other means than mechanical trituration available, by means of which thymol can be reduced to a much finer state of division than can ever be obtained by the agency of pestle and mortar. Though so little soluble in water, thymol is almost indefinitely soluble in rectified spirit, and can be conveniently kept in spirituous solution for dispensing purposes. If some of this alcoholic solution be poured into a relatively large bulk of water (say, a drachm of spirituous solution into an ounce of water), a milky emulsion at once results, which consists of minute oily globules, not much larger than those present in milk, and it is in this form that I would strongly recommend the drug to be administered, if we would wish to secure uniform results from uniform doses. If, however, such an emulsion be allowed to stand for a while, the circular, oily globules run together with larger ones, and, after attaining a certain size, resume the crystalline form, and then, though the crystals are usually but small, the advantage of over trituration is only one of degree. The addition to the water of mucilage or rice *conji* will retard this change for some time, but not sufficiently long to admit of the drug being sent out to patients in this form. Hence, the alcoholic solution should, wherever practicable, be only added to the emulsifying vehicle immediately before administration, and where the necessity of sending out of doses to patients renders this impracticable, it should be made with hot *conji* of such strength as to gelatinize when cold.

I believe that 20 grains of thymol administered in this way will be at least as effectual as 30 given in the usual fashion. To go into practical details I would recommend the following formulæ :—

(a) Stock solution of thymol—

Thymol	grains 480
Sp. rect.	ounces ii.

Dissolve the thymol in the spirit, and add a sufficiency of rectified spirit to make up to ounces iii. Twenty grains of thymol are contained in each drachm of this solution.

(b) Vehicle for administration—

Strong mucilage of acacia or dextrine	ounces ii.
Glycerine	drachm one-half.
Water	add ounces ii.

At the time of administration add one drachm of the stock solution of thymola to ounces ii. of (*b*), shake well, and administer at once.

Another point in connection with the use of thymol which requires some notice is the concurrent administration of purgatives. It is usually recommended to commence by the administration of a purgative, and also to follow up the doses of thymol with a second. Thus, treatment of a very active character, consisting in all of five doses of very active drugs, is extended over several hours, and the whole proceeding becomes so heroic that it becomes quite beyond the endurance of advanced cases, and, in some cases that have come under my observation, I am convinced has accelerated the fatal termination, though doubtless these were cases which must inevitably hàve died sooner or later whatever treatment had been adopted. I must confess that I am unable to understand the *rationale* of the first purgative dose. The intention is, I suppose, to obtain a free field for the thymol in the shape of an empty bowel, and, in the case of parasites whose habitat is the lower bowel, its administration would be rational enough. But the duodenum and jejunum, as the name of the latter indicates, never contain sufficient food to interfere with the action of the drug, and where we have to deal with *Dochmius duodenalis* its administration can only be needless, and will probably be prejudicial, for the irritating effect of the purgative will greatly increase the amount of mucus in the parts of the bowel over which it has passed, and this mucus will be a much more effectual bar to the action of the thymol than the presence of a considerable quantity of food. Besides, as far as these parts of the intestine are concerned, the withholding for a few hours of all solid food will be more effectual than all the purgatives in the pharmacopæa.

A previous abstinence from solid food forms part of the ordinary treatment, and I have been unable to trace the least additional advantage in the additional administration of a preparatory purgative in cases treated for comparison on the two plans. In the case of subjects in the initial stages of the disease, picked out by inspection, it is probably a matter of little moment which course is pursued, though I believe the preparatory purge to be in any cases needless, but, in the cases we are more commonly called upon to treat, especially in dispensary practice, every additional strain on the patient's strength, and especially on his digestive powers, becomes a matter of serious import, and, when diarrhœa is a prominent symptom of the fatal dyspepsia of the disease, it becomes a question whether the second purgative should not also be dispensed with, which, where the patient's forces are at all a low ebb, must be answered in the affirmative. As, however, has been already remarked, the action of the drug seems often to stupefy rather than kill the parasites, a subsequent purge is clearly indicated wherever the patient's strength will admit of it, lest the parasites should recover their vitality before they have been expelled from the intestinal canal. The danger of this is, however, perhaps more apparent than real, as, provided the narcotized parasites had been passed beyond the jejunum, the new habitat would be so unsuitable to them, that it is highly improbable that they would be able to establish themselves in these lower parts of the intestinal canal. Hence, wherever there is any doubt of the patient's ability to bear vigorous treatment, I should personally incline to the exhibition of thymol alone, except in cases where actual constipation is present, where the long retention of the drug in the system might lead to a larger absorption than is desirable. On the whole, as the mildest of all purgative agents, good castor-oil is probably the best drug for our purpose, and has only one disadvantage, in the fact that the presence of the oil in the dejecta renders the detection of the worms a matter of considerably greater difficulty than when saline purgatives have been employed. The way, indeed, in which it clings to the worms strongly leads one to suspect that its use as a preparatory purgative, so far from being advantageous, may defeat our object by giving the worms a protective coating, and so preventing contact with the thymol.

While deprecating the purposeless waste of thymol by the exhibition of absurdly small doses, it is to be noted that there are many indications that a dose of 30 grains is about as large as can be safely given, and that in advanced cases the practice of repeating this dose three times is not unattended with danger. In particular, I recall a case observed by Dr. Mullane and myself in Gauháti jail. The patient, who, while exhibiting unmistakable symptoms of anchylostomiasis, seemed by no means in the very last stages of the disease, was given thymol in the usual way. The worms were expelled in fair numbers, but the man died within 48 hours of the treatment. On *post mortem* examination, it was found that not a single worm had escaped the action of the drug, but that the whole length of the intestinal canal was in a state of active congestion, and there can be no practical doubt that the man's death was accelerated by the treatment adopted. Another reason for thinking that, in the customary dose of 90 grs. we are closely approaching the limits of toleration of the drug, is to be found in the fact that I certainly succeeded in poisoning a horse with a dose of $2\frac{1}{2}$ oz. This quantity was, it is true, given in a special manner, and in a single dose, but it is not much in excess of the customary dose for a man, proportionally to the weights of the subjects, especially when we remember that the whole 90 grains is usually in the intestine at once, no motion of the bowel commonly taking place until some time after the administration of the last 30 grains. I had some time previously given the animal $1\frac{1}{2}$ oz. in three $\frac{1}{2}$ oz. doses without producing any particular symptoms beyond the expulsion of a number of parasites, but it died within 6 hours of the administration of the larger dose. For about an hour after receiving it, the animal fed naturally, and seemed under no inconvenience, and I then went away, but, shortly after this, he became restless and weak, and laid down in his loose box and soon became unable to rise. He broke out into a "cold perspiration," and died apparently in a faint, having at no time exhibited much sign of pain. For the later symptoms I have to depend upon the description of Mr. Brown, the Transport Conductor here, who is, however, a very acute observer of animals. The animal seems to have died from cardiac failure, and the symptoms bear some resemblance to those of carbolic acid poisoning, as, indeed, might be anticipated would be the case from the chemical constitution of the drug. It should be added that the animal, a 13-2 Bhutia pony, though excessively weak and anæmic from the effects of *Sclerostomum tetracantham*, a parasite which extracts blood from the mucous membrane of the lower bowel of equine animals, in exactly the same way as is done for the human small intestine by *Dochmius duodenalis*, was by no means in a dying state, and had walked $1\frac{1}{2}$ miles just before without showing much distress. Any attempt, therefore, to increase the customary dosage should be made with the greatest caution.

The scantiness of my clinical opportunities prevented my making many experiments with other drugs than thymol. As to the efficiency of this drug there cannot be the least doubt. It expels round worms more surely than santonine, and, as I am told by Dr. Hancock of Dibrugarh, is far more efficacious than male fern in the treatment of tape-worm. I have myself seen it expel flukes, and there can be no doubt that it is an almost universal vermifuge. Hence the most important point was to, if possible, ascertain the best way of administering it, and the examination of other drugs was a, secondary matter. With salol, some success was obtained, but it appears inferior to thymol, and, as it is somewhat dearer, no advantage would be gained by its substitution. Picric acid had no effect whatever, neither, as far as concerned anchylostomes, had santonine. I meant next to have tried some of the less soluble analine compounds, but it would hardly be justifiable to try an experiment of this sort on any but an early case, and such cases I have not been able to obtain for treatment. Moreover, in the entire absence of information as to the safe dose of these substances, one would have to proceed very cautiously, so that a large number of tentative administrations would be necessary before one could be sure of one's ground. The idea that they might be useful is based on their known antiseptic powers; but, after all, the difficulty lies not in the expulsion of the parasites, but in the restoration of the digestive powers, and as no drugs can ever be expected to effect this, the whole art of treatment resolves itself into the early detection of cases.

COMPLICATIONS OF ANCHYLOSTOMIASIS.—*Other parasites.*— probably to the peculiarly favourable nature of the climate for the development of the free stages of nematodes and other parasites, Assam is a perfect hotbed for parasites, man and animals being alike affected, with an universality that can hardly be surpassed elsewhere.

It is excessively rare to find an Assamese entirely free from parasites, and, so far as my observations have gone, it seems almost equally rare to find it the case in an Assamese animal, whether free or domesticated.

In man, besides *Dochmius duodenalis*, the following parasites were met with :—

Trichocephalus dispar.— It is exceptional to meet with an Assamese who does not harbour this parasite. So far as my own observations go, it appears to be almost confined to the cœcum, and as many as a hundred may be present. It buries the long filiform aenterior, part of its body ·in the mucous membrane, never penetrating deeply, but burrowing along, barely beneath the epithelium, and subsists probably on the intercellular plasma of the tissues surrounding the tunnel it makes. Sometimes it does not seem to cause much irritation, but at others an angry blush marks out the tunnelled line ; and I have so often found it coiled up within ulcers of a dysenteric character that a causal connection between parasite and

K

ulcer may be strongly suspected. It, or a species anatomically identical, was present in all five of the specimens of *Macacus rhesus* that I have examined, so that the eggs must be very widely diffused over the country. Its presence does not appear to be incompatible with apparent health; but if, as appears probable, it be capable of causing ulcerations of the gut, it is evident that, under certain circumstances, it may by itself cause dangerous symptoms, and at best it cannot fail to do something to help the destruction of the digestive powers initiated by anchylostomiasis.

The life history of this parasite is involved in doubt, but the general opinion seems to be that infection is direct, *i.e.*, by ingestion of eggs. The structure of the envelopes of the ova are somewhat in favour of this view, the shell* being hard, and remarkably resistent to ordinary chemical agencies, a fact which leads one to suppose that they are designed to persist a considerable time as ova, and that some special agency, such as that of the gastric juice, is required to set them free. Leuckart* states that they require a temperature of not less than 72° F. for their development, and that the process is rarely completed under a year. When deposited, the ova, which are of deep red brown colour, of elongaed oval form, and possessing a peculiar sort of knot at each pole, show no signs of segmentation. In a cultivation started five months ago, on the same plan as those of *Dochmius* ova, segmentation of the ova commenced only five days after the commencement of the experiment. The yelk divides into two very unequal parts, and then the larger segment divides into two, and after 20 days, the morala stage was reached; but at present many of the eggs are still very backward, only a few contain embryos, and I cannot be certain that any have hatched out. The temperature has rarely exceeded Leuckart's minimum to any great extent, and has often fallen below it during the rains. A similar cultivation was kept for about a fortnight at a temperature ranging from 80°—104° F., but no particular hastening of the process took place during this period, and the temperature regulator worked so ill that a continuation of the experiment had to be abandoned. A practical difficulty in this investigation is that of getting specimens of dejecta containing *Trichocephalus* ova alone. On the other hand, certain cultivation, conducted in the open air and containing, *inter alia*, *Trichocephalus* ova, have been found, after a week or so, to contain

---

* Hoyle's Leuckart's " Parasites of Man," I., page 56.

a peculiar rhabditis, which is certainly referable to no stage of *Dochmius*, and in which the œsophagus runs through a peculiar cord of cells, the structure of which strongly recalls that of a similar structure found in the entozoic *Trichocephalus*, but, at the same time, other cultivations designed specially with the object of obtaining such rhabdites entirely failed to show any, so that the conditions, even supposing them to have been in these instances accidentally complied with, involve something not plainly apparent, possibly an invertebrate intermediate host. It is clear, however, that more time than as yet has elapsed is necessary for the thorough investigation of the life history of a parasite about which there is good authority for expecting a development extending over an entire year within the egg alone.

Trichocephali offer more resistance to thymol than any other parasite except *Oxyuris vermicularis*, but this is probably due only to the habitat of these parasites being so low down in the bowel. The drug is, however, far from being inoperative, so that it is common to find a dozen or more on the dejecta after the customary dosing, but, as a rule, as many or more will be left behind.

The prevention of infection, as indeed in the case of all intestinal parasites, is necessarily entirely a matter of conservancy. Moreover, certain observations I have made tend to show that the life of this entozoon is but short, and hence such measures alone will rapidly effect a cure by mere prevention of re-infection. Three little monkeys that I obtained for experimental purposes, five months ago, positively swarmed with *Trichocephali*, but now, simply by being cleanly kept, scarcely harbour any of them, and this can only be owing to the death and expulsion of the majority of the parasites, combined with the prevention of re-infection.

Ascaris lumbricoides is almost as common among the Assamese as *Trichocephalus*, and not unfrequently exists in such large numbers as by itself to constitute a distinct danger to life. I have removed 45 large worms from the body of a child a few years old, the whole mass weighing about ¾℔, or nearly one-fiftieth of the weight of the host. Although perhaps the commonest of all human parasites, we are in almost complete darkness as to its life history, though direct infection by ova is surmised. The large slightly elliptical ova have a very thick, nodulated envelope of a bright yellow colour, the density of which makes observation of the interior a difficult matter. When

K 2

deposited, there is no sign of segmentation, and I have never seen any change in such ova as I have kept under observation, but some accidental observations place me in exactly the same position as I stand with regard to *Trichocephalus*, as lately, in some old cultivations, some peculiarly large rhabdites have appeared entirely distinct from any stage of *Dochmius* and from the possibly *Trichocephalus* rhabdites.

This parasite, as is well known, often causes serious symptoms, and its extreme prevalence in Assam is no doubt one of the causes of the low average of physical health of the inhabitants.

In the treatment of patients for *Ascarides*, thymol has proved itself infinitely more sure and effectual than santonine, having often caused the expulsion of large numbers after santonine had been tried and had failed to expel more than two or three, so that, personally, I am unlikely ever to use santonine again.

Oxyuris vermicularis is not uncommon in Assam, but is proportionately much less so than the two preceding. The ova are thin-shelled, unsymmetrically oval, and always contain an advanced, lively embryo, which hatches out rapidly even in water. Direct infection by ova seems to be considered as proved, alike by Leuckart and Cobbold, but a species, anatomically indistinguishable, which infest *Macacus rhesus*, certainly goes through a distinct rhabditis free stage, and, though prevention of direct self-infection was clearly impossible, however cleanly the animals may be kept, they have, like the *Trichocephali* become distinctly less numerous during the period the animals have been in captivity. Of course, however alike, the human and simian parasites may be specifically distinct, but possibly both methods of infection are possible, and indeed the circumstance that oxyurids are common enough in European towns, where conservancy arrangements *qua* the propagation of intestinal worms, may be considered as almost perfect, may be alone considered sufficient to prove the possibility of self-infection. They are clinically unimportant in connection with anchylostomiasis, and are less affected by thymol, administered by the mouth, than any other parasite. A copious enema of weak emulsion is, however, far more effectual in expelling them than infusion of quassia, or any of the more commonly used remedies.

Amphistoma hominis, Lewis.—This curious little fluke which, curiously enough, was first found at Gauháti by Dr. J. O'Brien in 1857, and was described by the late Dr. T. R. Lewis in 1876[*] has been repeatedly met with during the present investigation, though not in numbers sufficient to constitute a danger to the patient. Its habitat is the cœcum and colon, and thymol appears to expel them with tolerable certainty. A tailless *cercaria*, possibly of this species, was, on one occasion, found in some mud used for plastering the walls near a hut inhabited by a subject of this parasite. Lewis and McConnell found them by hundreds, in which case they would necessarily develop a clinical significance, but personally I have never observed more than a dozen or so. I have never met with the ova in the dejecta even in cases which afterwards proved to harbour it, so that possibly these only reach the exterior on the death and expulsion of the parent.

Distoma Crassum, Busk.—A single specimen of this large fluke was expelled by thymol from a little girl at Chaygaon. Like many of her compatriots, this child harboured quite a museum of parasites in her own person, every one of the species hitherto mentioned being represented. So far as the scanty literature at my disposal enables me to judge, this is the first occasion in which the occurrence of this parasite in India has been recorded. In the case of a missionary and his wife who had worked at Ningpo in China, treated by Cobbold,[†] the most prominent symptoms were a distressing diarrhœa, hepatic derangement, and wasting, but very little is known about the species, and its habitat is only surmised to be the hepatic ducts, and the only instance in which it came under my observation being unexpected, and complicated by the presence of so many other parasites, I am unable to do more than to direct attention to its occasional occurrence in Assam, and the possibility of its becoming the cause of symptoms not altogether unlike those of anchylostomiasis.

Cestode parasites.—Owing to the circumstance that the people amongst whom most of my observations have been made, never eat beef, and rarely consume pork, I have personally never met with a case of tape-worm complicating anchylostomiasis, but I am told by Dr. Hancock of Dibrugarh that he has often met with, such cases among tea-garden coolies, and I understand that these worms are very common amongst the wild hill races on the south-eastern frontier of the Province. The species would almost certainly be *Tænia*

---

[*] Lewis McConnell. Proc. Asiat. Soc. Bengal, August, 1876.
[†] *Op. cit.*, page 22.

solium, and, from what I can gather from others, thymol would appear to be a far more efficient vermifuge in dealing with them than male fern or any of the other drugs more commonly employed for the purpose.

Pediculi.—In the course of examinations of the dejecta of patients affected with anchylostomiasis, I have repeatedly met with large numbers of immature *pediculi* crawling about among the fæces. For a long time I believed their occurrence to be purely accidental, and that they had gained access to the stools after deposition ; although in some instances the rapidity of their appearance made this, in the case of a wingless insect, extremely difficult of explanation. Quite lately, while examining an unopened duodenum preserved in spirit, I found six or eight of these insects actually in the bowel, and there can now be no doubt that the specimens met with in dejecta were deposited with them, or, in other words, that the *pediculi*, for some short period of their lives, had actually been entozoa. Now, this is somewhat remarkable, as there is no previously known instance of any non-larval insect acting in this way.

I sent down one of the most advanced specimens I could find to Mr. J. Wood Mason, the Superintendent of the Indian Museum ; but, owing to its immaturity, he tells me that he is unable to identify it with any known species.

I do not for a moment suppose that these insects pass any length of time in the bowel, but, at the same time, believe that they were born there. Any one who has watched the unsophisticated Assamese peasant when at his ease must have noticed the habit they have of searching each other's heads for vermin, and I have more than once seen the searcher deliberately swallow the prey he had captured. Doubtless, the adult insect gets crushed by the fingers and teeth ; but the ova would be unharmed, and, stimulated by the warmth of the intestine, would hatch out rapidly. In no other way indeed can the presence of these insects living, and in considerable numbers, in fresh dejecta be accounted for, especially as all are still immature when met with. Since making this discovery, I have had no opportunity of investigating what species of *Pediculus* infest the persons of Assamese villagers, so that I am still in doubt as to what species these temporarily entozoic insects should be referred.

They probably have no clinical significance, and I merely here note their occurrence, because it is desirable to put on record a fact which is, I believe, new to medicine.

Intestinal Coccidia.—The frequency of the occurrence of these organisms has already been noted on a previous page, and needs no further notice here.

It must be confessed that the above is a formidable list, and when the great frequency of many of them is remembered, it cannot be doubted that amongst them they form most efficient aids to the action of the arch-offender *Dochmius duodenalis.*

Malarial Complications.—Complications of this character are inevitable in any and every disease met with in such a climate as Assam; but at the same time I believe that its importance as a factor in the present increased mortality has been enormously overrated.

A very good idea of the part it plays in an outbreak of *kála azár* may be gathered from the tables given of the relative prevalence of this complication in the case of Chunsali tea-garden (*vide* page 143), but so much has already been said on this subject in different parts of this report that its further discussion here is unnecessary.

The Mode of Death in Anchylostomiasis.—Broadly speaking, death in this disease always occurs from asthenia, but the actual termination is usually brought about by the supervention of some more or less acute complication.

Of these, perhaps the commonest of all is dysentery, which seems occasionally determined by the action of *Trichocephali.*

The frequency of phthisis has already been noted by Leitz, and several cases of the kind have come under my own observation. More frequent than this, however, are subacute broncho-pneumonia and bronchitis, which, acting on organs already almost incapacitated by œdema, are rarely entirely absent, and rival bowel complaints in the frequency with which they determine the fatal result.

The occurrence of other complications has been noted, but they may be considered as rather accidental than as forming a distinct feature of the clinical picture; the above complications being present either together or alone in by far the larger proportion of cases, while death from the disease itself, without determining complication, is certainly the rarest incident of all. One such case, however, occurred in the Gauháti dispensary, brief notes of which have already been given, and here the patient died in the most typically asthenic way it has been my lot to witness, the temperature falling, and the lungs and heart hampered by dropsical effusions, acting more and more weakly, until the functions of life came to a stop, much in the same way that a machine is brought to a standstill by the dying out of the fire used to originate its energy.

X.—PREVENTIVE AND REMEDIAL MEASURES.

While this is undoubtedly the most important section of this report, the conclusions to be drawn from the facts recorded in previous sections, are so obvious that it appears almost superfluous to record them. It has been shown that, under different names, alike in native villages and among tea-garden labourers, anchylostomiasis has for a long period been causing a terrible mortality, and that nothing that has as yet been done has had any effect in checking it. It has further been shown that the disease is spread entirely by the agency of the dejecta of the patients: that, apart from the improvement of the condition of a few individuals, nothing can be hoped for from medical treatment; and that the improvement of even these few is unlikely to be permanent, as long as the material of infection is allowed to lie broadcast over the land. From these facts it follows absolutely that there is only one possible way of dealing with the disease, and that is to no longer allow the general fouling of inhabited sites, in other words, to adopt ordinary measures of conservancy.*

It is perfectly clear that even under the horrible old English cesspit system, the spread of this disease would be a physical impossibility, and that even the most rudimentary application of conservancy will go further to diminish the disease than all other measures, sanitary and curative, put together.

In the case of tea-gardens I can see no possible difficulty in the matter, but in Assamese villages I am well aware that the introduction of sanitation is beset with difficulties quite apart from the question of expense, and hence I could wish it were my lot to have to recommend measures easier of realization. This is, however, clearly impossible. To introduce a good water-supply and drain village sites are no doubt desirable, and would necessarily go far to improve general health, but they can do nothing to check the spread of anchylostomias, nor have they done so where attempted. Even supposing drinking-water to be the common vehicle of infection, and strong grounds have been given for believing it to be the rarest of all, to attempt to keep it pure by masonry wells, service tanks, and so on, while pollution of the soil around is allowed to any extent, is obviously a circuitous and inefficient plan of going to work, and leaves us much in the position of

---

* *See* paragraph 2 of Circular No. 14S., dated the 10th May 1889, and also paragraph 8, Sanitary Commissioner's Circular No. 1S., of 25th March 1890.

one who, wishing to prevent accidental poisoning, kept his water in carefully corked bottles, while allowing arsenic to be strewed about anywhere.

We are thus unfortunately placed under the dilemma of either introducing conservancy, or of leaving the disease to settle the question by depopulation. The introduction of conservancy must necessarily be very incomplete at first, but this should not prevent our at least attempting it, for, even if imperfect, it cannot fail to do something to improve existing conditions, for the burial of every single infected mass is a distinct blow at the spread of the disease.

Such measures have been successfully introduced in other parts of India, and though it is the fashion to speak of Assam as an exceptionally backward province, it must be acknowledged that many other parts of the country are not one whit more advanced, and though the Assamese are perhaps a shade more dirty in their habits than the average Indian, the difference is too small to involve any great increase in the difficulties of the problem, while, in general civilization, I fail to see that they are at all behind their neighbours, outside the larger towns.

The question of village sanitation is cropping up all over India; but, owing to the presence of this terrible disease, it is nowhere so urgent as in Assam, and, save by its introduction, I fail to see any hope for the amelioration of the disease. The exact plan of operations and the methods of securing their adoption are outside the province of the sanitarian, and all that need be done here is to indicate what appear to be the simplest and least expensive methods of conservancy, assuming their adoption to be enforced.

There is one fact which has been discovered in the course of the present investigation, which is of immense practical importance in this connection, and that is that the infective embryos can only live on the surface of the soil. Bury them, and they are rapidly killed, and hence any simple system of trench latrines will be quite sufficient for our purpose. The great practical difficulty in the matter is the scattered character of the Assamese villages, owing to which a considerable number of small latrines will be required to meet the convenience of the people; indeed, it is to be regretted that we cannot adopt the system enforced by the Jewish legislator (Deuteronomy, Chapter XXIII., verses 12-13), but, this being impracticable, the number will have to be large in proportion to the population, as, unless they be made a convenience instead of a burden to the people, they are unlikely to be used. In most places the expense of standing latrines would be prohibitory, but simple trenches might be adopted, into which a little earth should be daily thrown until they are nearly filled, when they should be closed and a fresh trench dug. Standing latrines furnished with tarred gumlahs and regular washing platforms would of course be better, and might be adopted with advantage on tea-gardens and in towns, but they would be too expensive for villages, and the want of a sufficient population of the sweeper caste would be an almost insuperable bar to their use, as regular removal of the sewage to a trenching-ground is absolutely necessary. There is nothing, however, in either digging a trench, or hoeing a little earth into it, which need hurt any man's caste, and the whole expense of such a system would be very small.

No doubt, in the rainy season, the trenches would get full of water, but this would be a matter of little moment, as the ova and embryos would sink to the bottom, and they would get buried just as surely, when the trench was filled in, as if it were dry.

A proper dry-earth system would undoubtedly be far superior, but, as the attainment of this is probably beyond village finances, we must be content with such simple measures as are within their means, remembering that, in dealing with anchylostomiasis, our one object should be to bury human filth out of sight, and that any measure which will even confine defilement to certain selected areas is better than nothing. In large places, such as Barpeta, such a plan would be out of the question. For such populations the adoption of proper dry-earth latrines, with a proper conservancy staff, is the only possible way of dealing with the matter; but, in such places, the expense would not be prohibitory, for the material prosperity of the inhabitants of Assam is certainly greater than those, for example, of the Central Provinces, and there, I know from experience, that a very fairly efficient conservancy is carried out by even small municipalities of only a few thousand inhabitants. During my stay there, a return of inhabitants capable of paying for medical attendance at the dispensaries was called for, and the extremely small numbers returned showed that the inhabitants were far from rich, so that the expense cannot be prohibitory in places, for example, of the size of Barpeta. In these towns in the Central Provinces the latrines formed a circle round the town, but were placed quite close to the houses. Each consisted of three buildings.

One designed for the use of males, and divided up into separate compartments, and another, of equal size, but not sub-divided, for females. Each was provided with a well-tarred platform for ablution, the drainings from which ran into a tarred receptacle, and water was obtainable close by. The third building was a shed for dry-earth. The buildings were constructed either of galvanized iron or of ordinary country materials, and the seats consisted of a couple of bricks, between which were placed two small tarred earthen *gumlahs*. In the male latrine the partitions were of ordinary bamboo matting. Some of the larger towns had proper conservancy carts, but, in the smaller places, the filth was removed in small receptacles by hand.

The conservancy staff was extremely small, so small, indeed, that it was surprising that the latrines were as well looked after as they were; it certainly did not exceed two men for every thousand inhabitants, and the fact that the sweepers somehow managed to do the work made me suspect that the latrines could be but little used. That this was not the case, however, I satisfied myself by personal observation. Some of the larger towns badly required additional latrines in more central situations; and I have no doubt that much defilement remained untouched, but the state of things was at least an immense improvement on the conditions that must have existed before their introduction, and nothing could show this better than the comparative rareness of parasitic disease, on the prevalence of which special returns were called for, the returns from the various dispensaries showing that very few such cases came under treatment.

An objection that is always raised to the adoption of the dry-earth system in Assamese towns is the difficulty of providing dry-earth. It is, however, purely imaginary. Assam, if one of the dampest, is not the only rainy country in India; and the difficulty is got over elsewhere by the simple method of storing a sufficiency of earth during the dry weather to last over the rains. In the part of the country I have just alluded to, during three months, the rainfall is quite Assamese in heaviness, but, though the sheds were very small, not 20 feet square, I should say, they contained a sufficiency of earth to tide matters over the rainy period, and, in this country, it can only be a question of increased accommodation, nor need be of an expensive character. Regular sheds are quite needless, beyond a small service shed, which may be located in a corner of the latrine, capable of holding a supply for a few days. For the main supply, all that need be done is to throw up a heap or heaps of earth during the dry weather, arranging it, so as to form it into the shape of a roof. The earth is then simply roughly thatched over with grass, and from time to time a sufficiency to fill up the service-shed is dug out from under the leeward end of the thatch. In this way, at the cost of a few rupees only, it would be perfectly easy to store an ample supply of earth to meet the exegencies of even an Assamese rainy season.

It is the duty of the sweepers to keep a little of the dry earth dusted into each *gumla*, and, on emptying these into the receptacles, to add a sufficiency to deodorize the mass. Once a day, the receptacles are removed, either in a cart or by hand, to the trenching ground, which should preferably be situated at a little distance from the town, but not at a sufficiently great one, to form a temptation to the conservancy men to shirk their work by emptying them into any hidden corner. In the absence of constant European supervision, this is a matter which is often not sufficiently taken into consideration. In view of the special dangers of anchylostomiasis, I should be inclined to recommend the use of deeper trenches than are usually considered desirable, say, $2\frac{1}{2}$ feet, and giving a full eighteen inches of clean earth on top of the soil, so as to bury all infective material thoroughly and deeply. Waste land should never be chosen for this purpose, as, after being used for trenching during one season, it is essential that the ground should be cropped; otherwise, in the course of time, it will lose its deodorizing powers. I have heard that an enormous yield of sugarcane can be got from land that has been so treated, but my own experience has been limited to corn lands, and, in one instance, to lucerne-grass, the yield of which was something astonishing. The drier the piece of ground chosen for the purpose, the better, but I do not look upon the trenches getting filled with water as an insuperable objection. The result will of course be not as good as in more favourable climates, but the essential point is to bury the infective matter, where it cannot be carried back into the houses; and, where the best is unobtainable, we must be content with the nearest approach to it we can get.

Disinfection of Sites already infected.—Having thus indicated the way in which the future spread of the disease may be surely prevented, it remains to be considered how sites, already infected, should be dealt with.

Provided only that all further supplies of nutritive matter be cut off by efficient measures of conservancy, it may be considered certain that, sooner or later, the infective embryos must die out, but unfortunately, we know that the time required for this may be very considerable.

It certainly exceeds six months, even in the absence of fresh nutritive supplies, and may be much longer. Given the continual supply of fresh nutriment, inevitable, where no attempt is made at conservancy, the infection of a site may last for an indefinite time, and indeed, is more likely to increase than decrease.

Two principal plans suggest themselves, the first of which has been found to be tolerably efficacious by the people themselves, *viz.*, desertion of the infected locality. Now, provided that the elimination of all infected persons could be secured by efficient medical inspection, this plan would be absolutely effectual even in the absence of all conservancy at the new site. Unfortunately, however, such a course is quite out of the question in actual practice, and hence migration only checks, and cannot stop, the spread of the disease, unless the inevitable infection of the soil of the new site be prevented by a proper conservancy. Without this, unless the people were to become absolute nomads, moving every week or so, migration can only be a palliative measure.

Given, however, the avoidance of fouling the soil of the new site, migration from infected localities may be strongly recommended wherever practicable.

The natural habits of the Gáros lend themselves easily to this plan of action, and, as the introduction of sanitary measures will be peculiarly difficult among such a wild race, they should be encouraged in this custom as far as possible. In the case of tea-gardens, the alteration of the site of the lines can often be carried out, and should be effected wherever practicable, of course coincidently with the introduction of proper measures of conservancy on the new site.

The disinfection of already-infected sites is doubtless a much more difficult matter, but cannot be regarded as hopeless. On account of the broadcast way in which the infective embryos are spread about such localities, and the impossibility often of recognizing the places where they are to be found, either by sight or smell, the use of all chemical disinfectants is clearly out of the question, but there are three facts (which appear in Section VIII.) of which advantage may be taken ; and these are—

1st.—Prolonged exposure to the direct rays of the sun kills the embryos.

2nd.—The same result may be effected by exposing them to a temperature exceeding 140°F.

3rd.—Burying the embryos kills them.

In order to take advantage of the first fact, all that is required is to clear the ground of vegetation as thoroughly as possible, at the commencement of the dry weather. The second property may be taken advantage of by firing the dry vegetation at the end of that season. For, as the embryos are confined absolutely to the extreme surface layer or the soil, the passage of a grass fire would certainly kill any that might have survived the dessication of the dry season. Lastly, by ploughing the ground at the commencement of the wet season, and so turning the surface layer under and burying the contained embryos, we may complete the disinfection, more especially of such parts of the sites as, for any reason, could not be treated either by clearing or by fire. This last expedient is, of course, available at any time of the year when the soil is soft enough for the process.

I feel assured that, by attention to these points, the disinfection of sites may be secured, though I necessarily have not been able to test them in actual practice.

Purification of water-supply.—It is a fortunate circumstance that the physical properties of the embryos are such as to render infection by drinking-water an unlikely contingency. They sink so rapidly in water, that, a short time after the introduction of even a large number into a well, all of them would become entangled on the mud at the bottom, and infection could only occur by the scooping up a portion of the sediment.

At the same time, more especially for the protection of officers travelling in affected districts, it may be well to point out that filtration affords an absolute security against infection by this agency. Any of the ordinary filters in common use are amply adequate for the purpose, as, owing to the, microscopically speaking, large size of the infective embryos, even the coarsest filters will exclude all chance of infection. In illustration of this may be instanced an experiment which I made, to ascertain how far an ordinary handkerchief would serve as a filter.

A piece of nainsook was placed over a glass funnel, and in the bag so formed was placed a quantity of earth which literally swarmed with embryoes. Water was then allowed to run through for a considerable time, and the filtrate collected. It was of course extremely turbid and obviously unfit for drinking purposes, but the most careful examination revealed only two embryos in the filtrate, while the residue on the nainsook was proportionately richer in them than ever.

With respect to attempts directed to the relief of the sufferers from the disease in Assamese villages by medical treatment, I have already plainly expressed my opinion. The attitude of the peasantry towards European medical treatment is such that it is perfectly certain that none but advanced cases will be persuaded to submit themselves to treatment. Such cases are practically incurable, and hence it is perfectly certain that the good effected by itinerant medical subordinates, extra dispensaries, and medical agencies in general, will be so minute that, for all practical purposes, any money so expended will be no better than wasted.

As an adjunct to proper sanitation, such measures can do no harm, and may even effect some small good ; but, unless proper measures of conservancy be adopted, the re-infection of the few cases cured will be so inevitable that nothing but discredit to the cause of European medicine can result.

In the case of the tea-gardens the case is quite different. The labourers, to begin with, come, for the most part from districts where the people have been long accustomed to the benefits of European medical treatment, and generally thoroughly appreciate it. Further, they are a great deal more under control, and can be dealt with, as far as sickness is concerned, almost in the same way as the men of a regiment, and, indeed the reception of all sick in regular hospitals is a well established custom.

With very few exceptions, however, treatment is carried out under the very mistaken policy of waiting till the man reports himself as sick, before attempting treatment. In the case of nine diseases out of ten, this is undoubtedly the best plan to adopt, but in that of anchylostomiasis it too often leads to the postponement of treatment until the man is beyond cure, and, at best, necessitates prolonged treatment and disability for work in place of two or three days' stay in hospital only. On this account,* I cannot too strongly urge the adoption of systematic periodical medical inspections of all labourers employed upon tea-estates, combined with the prompt treatment of all cases that may be detected, in addition to which, it is needless to say that all newly-arrived hands should be inspected,

---

* *See* paragraph 49, Annual Sanitary Report for Assam, 1888.

and, if necessary, treated before being allowed to mix with the
others. At the same time, here also I must repeat that such
measures can only be considered as adjuncts to the adoption of
proper sanitary precautions, and, apart from these, can never
be expected to really stamp out the disease, as such measures
infallibly must do, if only carried out with sufficient efficiency.

The measures I have to recommend may therefore be
stated as follows—

A.—With respect to "*kála-azár*" in villages :—

> *1st.*—The adoption and enforcement of some simple
> system of conservancy.
>
> *2nd.*—Migration from infected sites when practicable ;
> the disinfection of infected sites, by the measures
> above described, when migration is impossible.
>
> *3rd.*—The improvement of water-supply where this is at
> present unsatisfactory.
>
> *4th.*—The clearing and drainage of ground included in
> village sites wherever practicable.

B.—With respect to "*beri beri*" in tea-gardens :—

> *1st.*—The enforcement of proper conservancy, preferably
> on the dry-earth system.
>
> *2nd.*—The removal of lines to a new site wherever the
> disease has been very prevalent.
>
> *3rd.*—Where this is impracticable or undesirable, the
> disinfection of the site by the means above
> indicated.
>
> *4th.*—The adoption of raised " chang " huts, such as are
> customary in Burmese villages, whenever new
> huts are constructed.
>
> *5th.*—The entire prohibition of ·small enclosures round
> or near huts.
>
> *6th.*—The prohibition of the keeping of cattle within the
> inhabited limits of coolie-lines, and the provision
> of suitable separate enclosures for the purpose.
>
> *7th.*—The provision of a proper water-supply in cases
> when this is at present unsatisfactory ; the points
> to be aimed at being—
>
> > (*a*) That the well or other source of supply should
> > be some little distance from all habitations.
> >
> > (*b*) That if a well, it should be completely protected
> > by a locked cover, so that water can only
> > be got by pumping.

L

(c) That it should be carried by means of pipes to tanks above level of soil placed conveniently among the dwellings.

(d) In certain cases, *e.g.*, when river-water is the source of supply, it will be necessary, and in all cases it is desirable, to filter it by some apparatus on the ascension principle.

8th.—The provision of a proper supply of green vegetables, either by cultivation and sale to the coolies at cost prices, or by the provision of patches of ground somewhat on the English village " allotment ground " system.

XI.—Anchylostomiasis in Tea-Gardens.

That the sickliness among labourers imported from India into Assam, has, almost from the first, been a matter of serious concern to the Local Government is a matter which may be gathered from successive emigration reports. It is of course only of late years that the dependence of the up-to-then inexplicable mortality has been traced to anchylostomiasis, but the connection is now so firmly established that it would be quite superfluous for me to enter upon any attempt to prove what is already admitted.

Practically speaking, where any garden remains for any length of time on the " black list," it may be taken as certain that the cases that make up the excess over the normal mortality will be cases of anchylostomiasis. Of course, an outbreak of cholera may put a garden on the list for any given year, but an establishment will be singularly unlucky if such an accident should occur on two or more successive years, and putting aside cholera, the cause of relegation to the "black list " will be found to be anchylostomiasis in more than nine cases out of ten. Practically speaking, there are no statistics available that can give even an approximation to the amount of mortality that should be credited to the cause. Of late years, an attempt has been made to arrive at it by introducing separate headings for anæmia and dropsy* in the mortality returns of unhealthy gardens, but the results, so far, are quite valueless, as often not a single case will be recorded in a whole year from all the unhealthy gardens of an entire district, and that in parts of the country where all European medical officers are perfectly agreed that the disease is extremely rife. The fact is that, with but few exceptions, the persons responsible for the diagnoses of the cases from which the statistics are compiled are the native doctors of the tea-estates. I may have been unfortunate in those I met, but, speaking only of those I have seen, I can only say that I have never met with a body of men more grossly ignorant of their work.†
In spite of the fact that anchylostomiasis has now for several years been known to be the main cause of mortality among tea-garden labourers, very few seemed to have the least power of recognizing cases, and their entire diagnostic powers seemed to

---

* Sanitary Commissioner's Circular No. 14S., dated the 10th May 1889, directed the term " anchylostomiasis " only to be used in all garden returns.
† See Rule 122 under the Inland Emigration Act, I. of 1882.

consist in the ability to make a lame translation into English medical terms of what the patient thought was the matter with him. To instance one case, met with among many, I may mention a black-listed tea-garden that I visited in company with Dr. Bannerji in the Sibságar district. On visiting the hospital, we asked the native doctor whether he had any cases of anæmia, and he replied that they never had had any. In spite of this, there were sitting in the hospital verandah some half-a-dozen advanced cases of anchylostomiasis, which had been entered in the register as fever, dysentery, bronchitis, spleen, and so on, according to the coolie's notion of what was the matter with him. I then asked them to assemble all they could of the working hands, and some 60 were got together. Of these, a full third were found to be suffering from the earlier stages of the disease. Dr. Bannerji was at great pains to try and explain to the man the symptoms and treatment of such cases, but, from the expression of his face, I am strongly inclined to doubt if he was left a penny the wiser, the fact being that the idea of finding out a patient's disease for oneself was something entirely outside his notions of medical practice.

However, while these special statistics are so collected as to be quite worthless, those of the general mortality upon tea-gardens are probably more accurate than anything else of the kind in India. Whatever may be the proportion of the total that is due to anchylostomiasis, it is certainly so large that any increase or diminution in the disease would exert a marked influence on the total mortality ; and, judging from the returns of the last eight years, there can have been but little change in the yearly number of victims, but what change there has been is for the worse ; the past year showing a higher total mortality than any of the eight except 1884 ; and, when it is remembered that this increase has taken place in spite of much general sanitary improvement, but little doubt can arise as to the steady increase of this disease, which, as we have seen, is untouched by any sanitary measures save one, which has been left neglected. The actual total mortality ranges from 36·2 in 1889 to 43·2 in 1884, and last year stood at 41·8. Considering that this mortality includes not only that of anchylostomiasis, but also of heavy annual outbreaks of cholera, the general conditions of coolie life on the gardens cannot be considered at all unfavourable, and if we could but stamp out anchylostomiasis, would undoubtedly compare well with those of any part of India. It is curious in looking over byegone reports to see that opinion has passed through exactly the same phases in the base of coolie anœmia as it is now doing in that of *kála-azár*. Previously to the recognition of the parasitic causation of the disease, we find exactly the same arguments produced as to its malarial origin, and the same wearisome invocation of " climate " as the cause of everything connected with health, not easily explicable. Naturally, too, there was at first some opposition to the renunciation of these time-honoured notions, such as, I have no doubt, will meet the recognition of the same fact in *kála-azár*, but now, in the case of coolie anœmia, very few traces of the opposition are to be found, as European tea-garden practitioners, with very few exceptions, have fully recognized the nature of the disease. Another statistical fact, that points even more strongly to a steady increase in the disease, is the rapid rise in the number of gardens relegated to the " black list," or, in other words, with a mortality of over 70 *per annum per mille*. In 1885, there were in the Assam Valley but 23 gardens in which the mortality reached so alarming a height; in 1886, it had risen to 41 ; in 1887, the number fell a little; to 37, only to rise again, in 1888 to 54; and in 1889 to 88. No doubt the number of tea establishments has increased during the period in question, though I am unable to discover to what extent; but it is perfectly certain that the increase, whatever it may be, is insignificant in comparison with that of the number of black-listed gardens, which has nearly quadrupled during the short period in question, and at present 68,010 coolies, out of the 386·166 employed during the second half of last year, were working on black-listed gardens, or very nearly a sixth of the whole number.

In individual gardens the mortality is sometimes appalling, vying with that of the worst *kála-azár* villages, close on a third of the entire strength of a garden in one instances having died off in a single year, and instances of mortalities of over 150 per annum per mille are by no means uncommon, as may be seen from the subjoined table, which is compiled from the statistics published in the immigration reports.

Table showing the regular increase in the number of unhealthy gardens and their rate of mortality per thousand during the past five years.

Name of district.	1885.			1886.			1887.			1888.			1889.		
	Number of unhealthy gardens.	Average mortality in them.	Highest mortality in any one garden.	Number of unhealthy gardens.	Average mortality in them.	Highest mortality in any one garden.	Number of unhealthy gardens.	Average mortality in them.	Highest mortality in any one garden.	Number of unhealthy gardens.	Average mortality in them.	Highest mortality in any one garden.	Number of unhealthy gardens.	Average mortality in them.	Highest mortality in any one garden.
Kámrúp ...	3	104·6	188	Nil.	Nil.	Nil.	Nil.	Nil.	Nil.	2	81·5	86·3	1	96·0	96·0
Darrang ...	9	101·2	222	8	106·8	163·4	10	106·8	228·9	12	82·4	111·8	21	685	322·3
Sibságar ...	5	91·1	109·6	15	87·9	196·2	10	109·2	285	21	105·8	193·6	33	94·3	214·8
Lakhimpur ...	3	116·0	171	14	97·5	229·8	15	88·0	115·7	17	93·1	138·4	27	61·4	170·2
Nowgong ...	3	81·5	103·9	4	91·6	96·1	2	76·7	77	2	136	187·8	6	87·1	193·9
Cachar ...	1	...	...	10	84·8	229·7	7	91·6	114·0	·7	107·3	301	12	92·8	157·4
Sylhet ...	...	...	...	13	95·4	289	5	93·	117·2	4	85·0	101·4	10	95·1	181·4

Serious as it is, the mortality gives but a faint idea of the prevalence of the disease, the number actually affected being always largely in excess of the sick list. As an example, I may instance a special examination of Chunsali garden, near Gauháti, which was made on August 31st, 1890, with the following results :—

The estate finds employment, as a rule, for over four hundred hands, out of which, at the time of my visit, only 121 (53 men and 68 women) were contract labourers, the remainder being obtained in the local labour market. The whole of the contract labourers alike, working or on the sick list, were carefully examined for the symptoms of anchylostomiasis, enlargement of the spleen, and malarial cachexia, with the following results. Out of the total number, 65 presented more or less marked symptoms of anchylostomiasis, or 53·7 per cent. These cases may be classified as below :—

Table showing prevalence of Anchylostomiasis among Contract Coolies at Chunsali tea-garden.

Sex.	Diagnosis doubtful.	Unmistakeable symptoms.	Advanced symptoms.	Total.
Males	3	14	17	34
Females	3	12	16	31
Total of both sexes ...	6	26	33	65

Those entered under the heading " Unmistakeable symptoms " include cases where there could be no reasonable doubt as to the diagnosis, but which would still be, in all probability, amenable to treatment with thymol. . On the other hand, those included under that of " Advanced symptoms " are cases where the prognosis would be bad, whatever treatment might be adopted.

With respect to the prevalence of enlargement of the spleen, a greater or less degree of it was present in 45 out of the total number, *i.e.*, 37·2 per cent.

These instances of splenic enlargement were classified as follows :—

Table showing number of cases of splenic enlargement among Contract Labourers on Chunsali Tea-garden.

Sex.	Slight en-largement.	Moderate en-largement.	Considerable enlargement.	Total.
Males	10	7	5	22
Females	10	3	10	23
Total of both sexes ...	20	10	15	45

No relationship whatever could be made out between the prevalence of the two conditions. Some of the worst instances of anchylostomiasis had no splenic enlargement whatever, while, on the other hand, the case presenting perhaps the largest spleen of any showed no signs of anchylostomiasis, and laughed heartily at me when I asked him how long he had been ill, saying that he did not know that he was ill.

There were five cases of malarial cachexia either presenting no signs of anchylostomiasis, or but slightly so complicated, and some of the cases included under this head had but slight splenic enlargement. The degree of splenic enlargement was estimated by the native doctor while I was examining for anchylostomiasis, and the figures can therefore be in no way influenced by my own preconceived notions as to its absolute unimportance in the causation of *kála-azár.* It is, however, at least certain that he has not underestimated the prevalence of splenic enlargement, as the list includes many cases of abdominal enlargements from ascites consequent on advanced anchylostomiasis, where, personally, I could find no physical evidence of enlargement of the spleen.

The coolie-lines at Chunsali are placed on an excellent site, the natural drainage of .which is perfect. The well-water is bad, but the manager states they have been disused for a long time, and that nothing is used but Brahmaputra water, and I have little doubt this is broadly the case, as the river is nearer the greater part of the coolie-lines than the wells are, and probably only the few coolies who live nearer the wells than the river use the former. No systematic examination of the non-contract labourers was practicable, but I saw enough to convince me that they are at least as seriously affected as the contract hands, and in exactly the same way.

The instance of this garden is interesting, as it shows how entirely what is called *kála-azár* depends on anchylostomiasis, the Civil Surgeon remarking that he could see no difference* between this outbreak and those of *kála-azár* in villages. Now, in this connection, the above facts speak too strongly for themselves to need any comment; but it is worthy of note that the proportion of cases affected with enlargement of the spleen corresponds closely with Dr. Dobson's observations on its prevalence among reputedly healthy children.

The irregular and apparently capricious distribution of the centres of epidemic intensity in any given neighbourhood also exactly recall what was found to be the case on the distribution of *kála-azár*, and is an additional proof of the identity of the maladies known under the two names. It will be observed too, that, in both instances, while the severity of the disease is rapidly increasing in severity in localized centres, and while the number of the centres is steadily increasing, the increased mortality, when absorbed into the data of a whole district, is not sufficient to make any very distinct mark on its vital statistics, and this is exactly what might be expected in the case of anchylostomiasis, which, from the nature of the mechanism of its spread, must essentially be always a disease characterized by very localized centres of great intensity. It is also clear that, in spite of thymol, and of such attempts at sanitary improvement as have been made up to the present, the number of such centres is steadily increasing, and that their severity shows no sign of diminishing. I will now describe what I observed of the physical surroundings of the garden labourers as far as sanitary conditions are concerned.

(a) *Housing.*—There is no condition that varies more widely than this does in different gardens. In some cases, especially on the larger concerns, a great deal of money has been spent in providing comfortable habitations, while in others the "lines" consist of the most wretched hovels imaginable. On the whole, however, in this, as indeed in other sanitary matters, there is a praiseworthy eagerness on the part of the planting community to do everything in their power for the welfare of their coolies; but great mistakes are often made, and money spent to no purpose, simply through an ignorance

---

* *Vide* page 17.

of sanitary matters, in which, be it said, they are by no means peculiar. As a rule, the " lines " consist of thatched huts with wattle-and-daub walls ; but in some cases corrugated-iron has, with great advantage, been substituted as a roofing material, and more permanent materials have been employed as a portion of the walls.

About the best pattern of huts I met with consisted of two rooms, the gables and two pillars placed midway along the sides, consisting of rubble in mortar masonry, while the sides were filled in with wattle and daub work (*ekra*). The roof was of corrugated iron, and was prolonged beyond the gable containing the door, so as to form a sort of verandah. One drawback of its construction was an absence of ventilation. The lateral walls were necessarily too low to make ventilation openings in them tolerable to the inmates, while the solid nature of the gables prevented the upper part of these being utilized for the purpose. Again the advantage of the superior weather-tightness of the door gable was neutralized by its protection by the verandah. On this account it would be better, and probably in the end cheaper, to make the lateral walls of masonry and the gables of *ekra*, in which case the peak of the gables can be left unplastered as to afford sufficient ventilation without annoyance to the inmates, a plan which I have seen in use in many gardens where the huts are walled entirely with *ekra*. If then, the roof were prolonged at both ends, the whole of the mud-plastered portion of the building would be protected from the weather, and a very durable hut would result. Another fault was that the back room had no exterior door, and so was intolerably dark and stuffy. Whatever plan be adopted, it is perfectly certain that no room should ever be constructed without a door communicating with the open air. The best point about the hut was the provision of a verandah, which is a matter of great importance in a climate like Assam. Coolies should, however, be never permitted to wall these in, as they are very fond of doing, as, if this be allowed, they are nothing better than a hindrance to ventilation. In many cases I found small garden compounds, rudely fenced in, attached to the huts. They rarely, however, showed any signs of cultivation, and the untidy hedges were nothing better than convenient hiding-places for rubbish and filth of all sorts. The encouragement of the cultivation of fresh vegetables is, no doubt, a most important matter ; but I strongly doubt if it really is encouraged in this way, as scarcely any of these little patches showed signs of cultivation. Another most pernicious

•custom, which is permitted even on some of the best-managed gardens, is the close association of cattle and human beings. The large number of coolies who own cows is of course an excellent sign of their well-being, but this is no reason for permitting cattle and human beings to share one dwelling.

In the better cases the coolie builds a small lean-to shed against his hut, but everyone knows the filthy state in which natives keep their cattle-sheds, and, in a damp climate like Assam, the foul drainage from the cattle-shed cannot fail to soak through into the sub-soil beneath the hut. It is to be doubted if this arrangement be not actually worse than the actual association of cattle and men, as it is probable that, in the latter case, a certain amount of daily cleaning would be attempted. Provision should be, of course, made for the housing of the cattle, but it should be in a quite distinct enclosure, near to, but not in, the lines, all the cattle being kept together very much in the fashion of the cattle "kraals" in use among the Zulus. Care should be also taken that this enclosure should be placed lower in the drainage line of the land, and, if possible, to leeward of the coolie-lines with the prevailing wind. The ground so occupied soon becomes richly manured, and the Zulus utilize this circumstance by periodically moving their kraals, and sowing the ground with maize and vegetables, of which they thus obtain immense crops. This practice has, too the additional advantage of thoroughly purifying the ground, though it is needless to remark that this is a consideration which in no way influences the Zulus, for though their habits are, in point of fact, far less unsanitary than those of Indians, this is due rather to the accident of circumstances than to any particular desire or liking for cleanliness. Perhaps the greatest fault in the huts of all the coolie-lines I have met with, and I cannot recall a single exception, is that they are never provided with any plinth worthy of the name, an omission of a most fatal character in a country like Assam, where the raising of dwellings above the damp soil is a matter of the first importance. Of course, the provision of raised plinths would be a matter of considerable expense, especially as, in such a climate as this, nothing short of three or four feet can be considered adequate. Another disadvantage is that the necessary excavations would leave the ground honey-combed with depressions, which would certainly become receptacles for rubbish, and in the rains would be always full of filthy stagnant water. The whole plan then, of any of the huts I have seen appears to me radically wrong and entirely unsuited to the climate. What surprises me is that no one has thought of building their lines on the plan of Burmese villages, in which the huts are universally raised upon piles, the floor, which is made of split bamboo, being thus elevated some three to five feet above the level of the ground; a rude form, in fact, of the "chang" bungalow in common use among Europeans in Assam. The plan cannot be an expensive one, as it is to be seen in the poorest houses in the most wretched fishing villages on the coast. Moreover, the materials are at least as cheap and abundant in Assam as they are in Burma. That they are in universal use among so notoriously lazy a race as the Burmans, shows that but little additional trouble or expense can be involved in their construction.

No walls or obstructions of any kind should be allowed to interfere with the free passage of air beneath the floor, and, if the ground beneath the huts be kept clean, no better form of hut could possibly be contrived. In climate, density of jungle, soil, and all essential conditions, Burma and Assam closely resemble each other, except that, owing probably to the inferiority of the cold weather of the former Province, it is, if anything, rather the more malarious of the two, a fact shown by the much better health enjoyed by Europeans in Assam over their fellow-countrymen resident in Burma, and I am inclined to impute much of the undoubted physical superiority of the Burmese over the Assamese to the sanitary advantages of the national style of village architecture. The people here, too, are adepts in the craft of working with the bamboo. The walls of their houses being often constructed on exactly the same plan as the floors of Burmese raised huts, so that there could be no difficulty whatever in obtaining the necessary skill of labour.

There cannot be the least doubt that, with the exception of the special measures for the prevention of anchylostomiasis, no sanitary measure can be expected to do more for the improvement of the health of tea-garden coolies than the adoption of a style of housing suited to the climate in place of the system at present in vogue, which is modelled on the plan of village construction in use in India Proper, and which is entirely unsuited to the very different climatic conditions of the Indo-Chinese Peninsula.

There is, of course, no necessity for any slavish adherence to anything but the principle of the Burmese plan, and the adoption of improved roofing material, as well as of masonry supports, would be undoubted improvements, and would pro-

bably result in a saving of expense in the long run ; at least
that is one of the reasons advanced for the adoption of galvanized iron by some of the garden managers I have met.

Provided that, as appears to be the case, expense need not
stand in the way, the adoption of an impervious roofing material
in place of thatch would be an immense advantage, for the
mass of continually wet, rotting thatch is, I am convinced, a frequent source of ill-health not only in native, but also in European dwellings.

(b) *Clothing.*—In this matter the coolies do not seem ill
off, at any rate, they seem to have as much as they care to
wear, though it is probable that the adoption of woollen jackets
during the chilly fogs of the cold weather would be an improvement. Recommendations to this effect occur frequently
in the inspection reports of many medical officers, but the difficulty in providing the clothing would be but a small one in comparison with that of inducing the coolies to wear it. While,
however, I doubt if adults suffer in this way, there can be no
doubt that the children are universally wretchedly underclad,
and there cannot be the least doubt that neglect of this kind
is an important factor in the heavy infant mortality. It cannot, however, be said that tea-garden labourers are any more
careless in this matter than natives of India generally, though
its consequences are probably more serious in Assam than
they are in other, and drier parts of the country. It is,
however, to little purpose discussing the subject, as, pernicious
as the custom undoubtedly is, it is too deeply rooted in native
habits for any improvement to be hoped for.

(c) *Food* consists mainly of rice, pulse of various
kinds, fish, milk, and vegetables. Its adequacy depends mainly
on the state of the labourer's finances, and, in the case of new
comers, is often none too liberal. Old hands, however, are generally fairly well off, as may be seen by the large numbers of
cattle and poultry often possessed by them. From what I
saw I am inclined to think that the coolies, when free from
disease, are mostly well nourished.

In the case of immigrants coming from parts of India
where corn forms the staple food, a really serious matter
is the enforced change to the insipid and comparatively unnu-
tritious rice. So much larger a bulk of rice is required to meet
the demands of the organization, that a wheat-eater must needs
contract a certain amount of chronic dilatation of the stomach
before he can accommodate a sufficiently full meal. Unfortunately, wheat is an exotic luxury in Assam, but facilities for its

purchase should certainly be afforded in the case of men coming from the corn districts of India, at any rate, for a sufficient time to enable them to accustom themselves gradually to the new diet; and, as rice is considered a great luxury in their native homes, they would not probably be long in doing this. This is a matter, however, which has been repeatedly adverted to in reports from various officers, and I think its importance should not be underrated.

In the matter of vegetables, the coolies are often but ill supplied. Their own efforts do not go much beyond the cultivation of a few pumpkins, a vegetable which, however valuable, is by no means the best of antiscorbutics, while many of the gardens are placed in portions of the country, so destitute of indigenous inhabitants, that nothing of the kind can be bought. Outside the medical profession, people are too apt to underrate the importance of a due supply of food of this sort. This is, no doubt, owing to the small proportion of nutriment they contain, the meagreness of which is so obvious that it needs no acquaintances with analytical tables to inform any one on the point. Green vegetables, however, contain a large amount of certain organic salts which are absolutely essential to the preservation of health, and which are not contained in grain, flesh, or the other solidly nutritious articles of diet.

Owing to this want, scurvy is by no means rare amongst the immigrant labourers, and, as a man in this condition will never be fit to do a good day's work, I am certain it would often pay well to systematically cultivate *sag*, onions, cabbages, turnips, radishes, plantains, and other suitable vegetables, even if it involved the diversion of a certain amount of labour from the regular work of the garden. For want of a proper appreciation of the importance of green vegetables, scurvy is common all over India, as indeed it was in England when in a similar stage of civilization, so that it is useless to trust to the coolies' own instincts to overcome any notable difficulties in this direction, though he will doubtless avail himself of a cheap and plentiful supply when obtainable.

The question of food-supply cannot be passed by without some allusion to the "hotel system," of which I have heard much, though I regret to say I had but few opportunities of seeing the system in full working order, the time of the year being unfavourable for this. Sufficient, however, was seen to convince me of its immense value, which mainly lies in tiding the new arrival over the time during which the strangeness of his surroundings prevents his being able to look properly after himself. The only criticism I have to make on the subject is that, if it is desired to get the full nutritive value out of the food supplied, more attention should be paid to its cooking. If properly cooked, so that the grains are plump but remain separate, rice is one of the most digestible of foods; but if, as is often the case, it be allowed to get into a stodgy mass, closely resembling bookbinder's paste, it becomes heavy and hard of digestion. In the same way, *dal*, when properly cooked, is one of the most nutritious and easily assimilable forms of nitrogenous food, so much so, that, under the name of " Revalenta Arabica," it has gained a great reputation in Europe in the treatment of dyspepsia. Yet this very same article of food, if imperfectly cooked, as it generally is amongst natives, becomes extremely irritating, and much of it passes through the intestinal canal unaltered, and so is lost, or worse than lost, for not only has it in no way contributed to the nutrition of the body, but it has caused positive harm by producing irritation. When properly cooked, *dal* should form a perfectly smooth uniform pulp, no sign of the component grains being left, and, when so prepared, it is unrivalled as a source of nitrogenous pabulum.

Attention to little points of this kind in cooking, may make all the difference between recovery and death in the case of patients after treatment with thymol for anchylostomiasis, who form a large proportion of the inmates of an " hotel." The digestive powers of these poor creatures are always reduced to the lowest ebb, and the mere expulsion of the parasites is absolutely useless, a mere waste of an expensive drug in fact, unless it be followed up by a supply of food, not merely plentiful, but also so prepared that it can be made use of by their utterly enfeebled digestive organs. Now rice and *dal* can both be made, with very little trouble, into food highly suitable for invalids, but, cooked as they commonly are by natives, especially when the operation is conducted upon at all a large scale, they are little better than the loading of the intestines with indigestible, and, at the best, inert matter.

If, however, we can once check the spread of anchylostomiasis, there will be little further need for " hotels," for, as long as he is in health, the coolie may be trusted to take care of himself in the matter of feeding. It is commonly enough asserted that coolies will half starve themselves from motives of economy, but I entirely disbelieve in the commonness of such cases. In nine cases out of ten, miserliness has nothing to do

with the matter ; and the reason of the man's not buying a sufficiency of food is that he is too ill to care for it, and, naturally
enough, will not waste his dwindling resources on the purchase
of food for which he has no appetite.

In quite the earlier stages of the disease, it is true patients
often exhibit an increased appetite, but this is only for a short
period, and, as the digestive powers become more and more
undermined, and the patient a confirmed dyspeptic, the appetite
becomes small and capricious. Arrangements should, however,
be made on all gardens for a cheap and plentiful supply of
green vegetables just as they are in the case of the other ne-
cessaries of life, as it is impossible for the coolie to supply
himself with what is unobtainable.

(d) *Water-supply.*—More attention has been devoted to
this matter than to any other sanitary point, and there are
large numbers of gardens where large sums have been spent
in the provision of wells, receiving tanks, and even filters, and
the supply is often excellent. This is often especially the case
on "unhealthy gardens" which will be more frequently found
to possess a good water-supply than such as have never been
relegated to the black list. The reason of this is, of course,
that a heavy sick list has directed attention to the necessity of
sanitary improvements, and, as might be expected, a marked
improvement in general health has resulted. The improvement, however, fails to take the garden off the black list, as it
fails to touch the most important factor of the mortality, for I
could not find any instance in which the improvement of water-
supply could be shown to have done much, if anything,
to diminish anchylostomiasis.

In those instances where improvement had resulted from
the carrying out of sanitary recommendations, removal of the
site of lines would generally be found to be one, and as I
believe the, effective agent in the improvement. No one can
be less disposed than the writer to underrate the advantages
of a pure water-supply; but, as has already been said, I do not
think its provision can do much for the prevention of the particular disease we have under consideration. Let it not, however,
be supposed that the water-supply of tea-gardens in general
is everywhere as it should be ; indeed, so far from this being
the case, a supply of good quality is almost restricted to those
gardens where the spur of a heavy mortality has led to strenuous efforts at better sanitation. A glance at the table given on
page 32 *et seq.*, will show that the drinking-water of many
gardens is horribly foul, and the introduction of a pure sup-

ply in such cases must be a crying necessity. Even where
money has been freely spent, mistakes due to want of knowledge
of a very special subject have not unfrequently resulted in a
doubtful, in place of a thoroughly reliable, supply. For
example, in one of the most progressive establishments in the
Lakhimpur district, at the time of my visit, they were busily
engaged in digging a well. It was being sunk through an extremely porous, sandy soil, and, owing to its shifting character,
the amount of labour that was required was enormous.
In nearly any other situation but where it was placed, it
might be expected, when completed, to yield excellent water ;
but, instead of putting it well away in the midst of the cultivation, they had been at great pains to find it a place in the
very heart of the coolie village, which accommodated many
hundred souls, having actually gone to the additional expense
of demolishing some buildings to find space for the enormous
conical excavation necessitated by the unstable character of
the soil. Placed as it is, and in such a soil, defilement with
the foul surface washings of the lines will be almost inevitable, as no masonry tube, however well made, can be
trusted to entirely exclude percolation for any great length
of time. It was intended that the well should be covered, and
provided with a cemented supply tank, and the additional
cost of an iron pipe to convey the water from a well, situated
well away from the habitations, to a tank or tanks in the lines
would have involved an additional outlay so trifling as to be
hardly worth considering, as a part of the large expense that
had already been incurred.

(e) *Conservancy.*—In this matter the arrangements in
use upon tea-gardens may be very briefly described. There are
none. The entire neglect of this most essential of all sanitary
necessities is the greatest blot to be found in the condition
of the tea-gardens, and is the essential cause of the spread
and persistence of anchylostomiasis, to say nothing of the evil
effect it must have on the general health. The existing state
of things cannot be described in too strong terms. Crowded
together, as the huts necessarily are, and, indeed, should be,
for I do not use the term in the sense of overcrowding, the
filthy state of these settlements, in even the best-managed
gardens, is revolting. If you ask what conservancy measures
are adopted, you will be told that the people " go to the
jungle," or " amongst the tea-bushes," but a visit to the localities pointed out will seldom result in the discovery of
anything particularly offensive to either the nose, or eye.

M

It is between and behind the huts, and in the little enclosures already alluded to, that nine-tenths of the offence will be found to be concentrated, and, for all practical purposes, the ground immediately round a man's hut forms his only latrine. When this is pointed out, you will be told that "it has been done by the women and children," but the scantiness of deposits elsewhere makes this improbable ; and, even if it were true, its source in no way diminishes the harmfulness of the practice. Besides, even supposing the men do resort to the tea-bushes, &c., as described, it is difficult to see how we are more than a degree better off. It is perfectly certain that they will not go far afield, and that they will habitually resort to nearly the same piece of ground, and wherever cases of anchylostomiasis are to be found amongst the coolies, the result will be the production of a nursery for rearing the infective embryos of this most fatal of parasites, under the most favourable conditions imaginable ; for the trampling about of a considerable number of persons on the small area so used will result in the broadcast diffusion of the embryos over its whole surface, and the ready supply of fresh nutriment will ensure their indefinite multiplication.

In addition to human pollution, the droppings of cattle, stable refuse, and rubbish of every kind commonly litter the ground, and are seldom cleared away ; and when this is done, the refuse is seldom removed to a sufficient distance, or burnt. In one large garden, I found, flanking the lines, a mass of accumulated sweepings some three to five feet high extending along the whole of one face of the settlement. It was not more than twenty feet distant from the nearest huts, and was placed rather above them, on the gentle slope on which the lines were built. The surface of this mass, which must have represented the accumulation of several years, was evidently the "jungle" and "tea-bushes" I had heard so much about, so that the nature of its drainage, which necessarily flowed into the lines, may be better imagined than described ; and yet in this garden the expenditure on good huts, and the provision of a good water-supply could only be described as liberal in the extreme.

Tea-garden sanitation has, in fact, been commenced at the wrong end, for water-supply might, with comparative impunity, have been left to take care of itself, provided measures had been taken to put a stop to its pollution by means of adequate measures of conservancy.*

---

* See Circular from Sanitary Commissioner, No. 1S., paras. 8, 7, 25th March 1890.

It would be wrong, however, to impute much blame to the planter for the existence of this state of things, for the cause lies in the naturally dirty habits of the coolie, and, unless he resort to illegal means, the planter has no power to enforce the observance of more decent habits. All that can be laid to the planter's door is that, by omitting to provide proper latrines, he has neglected to give his labourers the chance of becoming cleanly. The cause of this neglect is mainly a rooted belief that, even if provided, they would not be used. This belief is, however, based on purely theoretical considerations of the innate dirtiness of the coolie character, for I found no garden where they had tested the theory by actual experiment. The argument is a stock one, which has been made use of by the opponents of sanitary measures, wherever proper conservancy has been proposed in India, and has almost universally proved fallacious when put to the actual test. Where the introduction of latrines has proved a failure, it has almost always been owing to mistakes in the choice of site. If they be placed at a long distance from the habitations of those who are to use them, they are naturally enough never resorted to; but if so placed and arranged as to be an actual convenience, they will soon come into use, for a man soon learns to prefer to go to a well-kept latrine, on a dry site, with water, and a platform for ablution at hand, to wading a longer distance, through a sea of mud. Once the matter is fairly faced, the difficulties will, I believe, be found to be mainly imaginary. During my recent tour I made at least one convert to these views, and I am told that the success which has attended the introduction of a simple form of latrine I recommended to him, has greatly surprised the experimenter, as the coolies very soon took to availing themselves of the greater convenience of the arrangement placed at their disposal.

The disposal of other offensive matter, and rubbish, which also forms an essential of proper conservancy, is a less simple matter, as it is hard to make a convenience of the labour involved in keeping the lines in a cleanly state. If health is to be improved, however, it is certain some organization must be instituted for the purpose. The nature of such an organization is somewhat outside the province of the sanitary officer; but personally I would make the coolie himself responsible for the cleanliness of his surroundings, and would limit the function of the European manager to directing the attention of the authorities to cases of neglect of proper sanitary precautions. There is this much to be said for the labourer, that he has not the least idea that there is anything in his mode of life dangerous either to himself or his neighbours, and that the law as yet puts no veto on these habits, although they are far more destructive to life and comfort, than many matters about which it minutely concerns itself. There is, of course, not the least chance of being able to convince the coolie that his habits are a standing danger to himself and every one around him, but whatever their faults, Indians are at least a most law-abiding race, and if they were once made to see that breaches of the common decencies of sanitation are illegal, they would slowly come to acquiesce in the inevitable, and adopt more cleanly customs ; but without some amount of legal coercion it is difficult to see how any great improvement can be expected. Such being, then, the conditions under which the coolie lives, it is in no way surprising that anchylostomiasis should be rife amongst them. Given the introduction of a single case into a community so situated, the spread of the disease becomes almost inevitable, and the result is that very few gardens can now be said to be absolutely free from it ; and, unless strenuous preventive measures be instituted, the number of gardens on the "black list" must necessarily go on increasing, even more rapidly than it already has, until nearly the whole of the establishments of the province come under this melancholy category.

With this section, this report comes to a close. The investigation into the method of spread of the disease leads to but one conclusion, for it follows from the facts I have endeavoured to set forth, that the only ·commonsense method of prevention is to adopt the absolutely sure plan of destroying all chance of infection, by proper measures of conservancy. From no other plan can any ‑appreciable improvement be hoped for. A thoroughly efficient conservancy cannot fail to ultimately entirely stamp out the disease, and, whatever success may be achieved in practice will be exactly proportional to the efficiency of the measures of conservancy adopted.

GEO. M. GILES, M.B., F.R.C.S., SAN. SCI. CERT. UNIV. LOND.,
SURGEON I.M.S., ON SPECIAL DUTY, ASSAM.

Shillong, 1st October 1890.

ADDENDUM.

On pages 7-8 of the present report, reference is made to a note in Dr. Kynsey's pamphlet on the beri-beri of Ceylon, on a paper by Leichtenstern on anchylostomiasis, and to the inexplicable character of the abstract given. During a recent visit to Simla, I found that the office of the Surgeon-General with the Government of India is provided with a file of the "Deutsche Medicinische Wochenschrift," and I accordingly proceeded to search for the paper in question.

It is not surprising that I was unable in the first instance to obtain it, as the authority appears to be incorrectly given. The paper referred to is, I presume, one entitled "Einiges über Anchylostoma duodenale" Von Otto Leichtenstern, Deutsch. Med. Wochen., Vol. XIII., 1887, pages 565, 594, 620, 645, 669, 691, and 712.

Though modestly entitled "a few words" on the subject in question, the article, it will be seen, is a very long one, running through no less than seven issues of the periodical. The main bulk of the paper is taken up with a criticism on the work of Dr. Schultess, a translation of which, it may be remembered, appears in Dr. Kynsey's pamphlet; but I could find no such statement as that quoted thence in the text of this report, no mention being made of so extraordinary a method of multiplication as by the separation of the "segments" of a rhabditic nematode.

I am glad to find that Leichtenstern entirely agrees with my own views as to the significance of the so-called calcification of the embryos, looking upon the change as merely an indication of death, and many pages of the paper are occupied in controverting the views of Schultess on the point. We are also in accord as to the explanation of this non-development of ova within the human intestine being due to the absence of oxygen, Leichtenstern having established this matter beyond doubt by rigid experimental tests, showing that cultivations immersed in other gases fail to develop.

It is, however, quite obvious that, like his predecessors, he quite failed to follow the rhabdites to maturity, and so missed the true significance of the free stage.

This failure I believe may be traced to his view (page 669) that water is the natural habitat of the rhabdites, whereas, as has been demonstrated in the text, no complete development of these organisms can take place in that medium, their true habitat being not water, but fœcal matter.

Hence he has no suspicion that a regular sexual multiplication of the free stage is possible and normal. Hence, too, his laborious description of the so-called encapsuling and re-incapsuling of the embryos, which are, of course, nothing more than ordinary ecdyses, greatly delayed by the unsuitability of the conditions to which his embryos were exposed.

He speaks, too, of successful feeding experiments; but, as they are not given in detail, their true significance cannot be judged.

A further point on which our experiences are in entire agreement is the non-discovery of any worms encysted in the submucosa, and indeed the remarks I have made on pages 85-86 of this report might almost pass as a free translation of what he has written at page 692 of the "Deutsche Medicinische Wochenschrift." He does not, however, dispute the accuracy of the older observations of Bilhartz, Greisiger, and Grassi (1852-1854); but looks upon the worms they found encysted as having "wandered" from their proper habitat.

This explanation, however, fails to satisfy me, as it runs counter to the general phenomena of helminthiases, and I should still prefer to believe that our failure is rather to be traced to neither of us having met with suitable cases.

Of great interest, too, are the cases he gives (page 693, instance,) illustrative of the longevity of the parasite. In one instance he considers he has good evidence of their having lived for years within the human intestine, but it does not seem quite clear that the possibility of re-infection is entirely excluded, his argument being based on the assumption that opportunities for infection are afforded only by brick-fields. In another, however, of his cases, a duration of two years appears to be distinctly proved, as the patient passed that period in jail, subsequently to infection, and it may be fairly considered an impossibility that infection should take place within the walls of an European prison. One would like, however, to know the nature of the labour on which the prisoners were employed. I have met with similar cases in Assamese jails; but here the possibility of re-infection cannot be entirely excluded, as the prisoners are much employed on extramural labour, in the pursuance of which frequent opportunities of infection may occur.

SANAWAR: }
December 20th, 1890. } G. M. GILES.

A. S. Press (General) No. 79a—520—14 1.91.

ERRATA.

Page 54, line 27 from top, *for* "ilum" *read* "ileum."

,, 60, ,, 18 ,, ,, "jejunam" *read* "jejunum."

,, 76, ,, 7 from bottom, *for* "rhabdites" *read* "rhabditis."

,, 91, ,, 26 from top, *for* "caleoptera" *read* "coleoptera."

,, 98, ,, 3 ,, ,, "rhabdites" ,, "rhabditis."

,, 98, ,, 14 from bottom, *for* "there" *read* "their."

,, 117, ,, 20 from top, *for* "with" *read* "into."

,, 117, last line, *for* "add" *read* "up to."

,, 121, line 22 from top, *before* "probably" *insert* "owing."

,, 122, ,, 22 ,, *for* "elongaed" *read* "elongated"

,, 139, ,, 9 ,, ,, "or" *read* "of."

,, 140, ,, 10 from bottom, *for* "measures" ,, "measure."

,, 141, top line, *for* "base" *read* "case"

Explanation of Plates.

Plate II., figure 2, *for* "pre-sexual" *read* "pre-sexual."

,, IV., ,, 1, ,, "ilium" *read* "ileum."

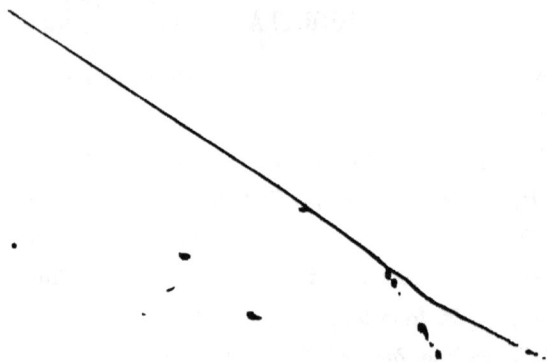